AF770611

BUILDING THE ATKINSON "CYCLE" ENGINE

Written and Illustrated by
Vincent R. Gingery

Published by
DAVID J. GINGERY PUBLISHING
P.O. Box 318
Rogersville, Mo. 65742-0318

Printed in the U.S.A.

FIRST EDITION

THIRD PRINTING 2008

COPYRIGHT © 1996 VINCENT R. GINGERY

ALL RIGHTS RESERVED

LIBRARY OF CONGRESS
CATALOG CARD NUMBER 96-75696

INTERNATIONAL STANDARD BOOK NUMBER
1-878087-18-5

TABLE OF CONTENTS

INTRODUCTION. 5

A BIT OF HISTORY 5

OPERATING PRINCIPLES 8

PREPARING TO BUILD 13

SOURCES FOR MATERIAL 15

MATERIAL LIST 15

HARDWARE 16

PRODUCING THE CASTINGS 17

MAKING THE PATTERNS 19

THE FLYWHEEL 20

THE BASE PATTERN 22

THE CONNECTING ROD PATTERN 23

THE PIVOT POST PATTERN 24

THE HEAD PATTERN. 25

THE CRANKSHAFT SUPPORT PILLAR PATTERN 26

THE PISTON ROD PATTERN 26

MAKE THIS A GROUP PROJECT 26

MACHINING THE PARTS & ASSEMBLING THE ENGINE 27

SPECIAL TOOLING 27

THE CRANKSHAFT SUPPORT PILLARS 29

THE BASE 31

MOUNTING THE CRANKSHAFT SUPPORT PILLARS 31

THE CRANKSHAFT & CRANK 34

MAKING THE CAMS 35

THE IGNITION POINT MOUNTING PLATE 39

INSTALLING THE CRANKSHAFT 40

MACHINING THE FLYWHEEL 40

MAKING THE CONNECTING ROD LINKAGE 42

THE WATER JACKET 47

BORING AND LAPPING THE CYLINDER 49

MAKING THE PISTON 54

ASSEMBLING THE LINKAGE 55

MACHINING THE HEAD 57

MAKING THE VALVES 59

ALIGNING THE LINKAGE 62

INSTALLING THE RINGS 65

THE ROCKER ARM ASSEMBLY 67

MAKING THE PUSHROD LINKAGE 67

MAKING THE CARBURETOR 76

MAKING THE GREASE CUPS 83

THE GAS TANK 83

THE CYLINDER OILER 86

THE SPARK PLUG. 86

THE IGNITION POINTS 86

THE ELECTRICAL SYSTEM. 87

SETTING THE CAMS AND ADJUSTING THE POINTS. 87

FINAL PREPARATION. 89

SPRUCING UP AND ADDING A FEW EXTRAS. 90

WARNING: THERE ARE SERIOUS HAZARDS IN THE PROCESSES AND PROCEDURES IN THIS BOOK. NO ATTEMPT HAS BEEN MADE TO POINT OUT ALL OF THE DANGERS OR EVEN A MAJORITY OF THEM. THE METHODS AND MATERIALS THAT ARE SUGGESTED IN THIS MANUAL WERE DEVELOPED BY A NON-PROFESSIONAL. THE AUTHOR IS NOT AN ENGINEER OR SCIENTIST AND NO CLAIM IS MADE TO THE PROPRIETY OF THE METHODS SUGGESTED IN THESE PAGES. THE READER IS FULLY RESPONSIBLE FOR DEVISING SAFE PROCEDURES FOR EVERY OPERATION.

INTRODUCTION

One of the most rewarding things to have come out of our series of books is the many friendships that have developed over the years.

Jim Lewis is one such friend. To say the least, Jim is an amazing man. His hobby is designing and building gas engines. And in that regard he is a genius. On a recent visit Jim showed my Dad and I some of his completed engine projects. Any one of them would have made an excellent subject for a book, but the one that caught our eye was an Atkinson "Cycle" engine he had built. Although his Atkinson engine is not an exact replica of the original built in 1886, its design principles are the same.

Figure 1 **This is a photo of Jim out in his shop showing us his Atkinson Engine.**

Jim came across the idea for his Atkinson engine while reading a book called, "Internal Fire" written by C. Lyle Cummins Jr.. While most of us would require detailed plans to begin such a project, Jim was able to build his engine using one simple drawing and a photo contained in the book "Internal fire". A photo of the engine as it was built in the late 1880's by Manlove, Alliott & Co., Nottingham is shown figure 2. For comparison, a photo of the engine built to Jim's design is shown in figure 4.

While watching Jim's engine run, we became fascinated with it and its unique design. It occurred to us that this engine would be a perfect and relatively simple one to build, not only for us, but for others. With this in mind, we approached Jim with the idea of building his Atkinson engine ourselves and drawing up plans for the project so that others could share in the joy of building such an engine. He was more than willing and allowed us to borrow his engine and casting patterns to use as an aid in building our own. The book you hold in your hand is the result. Jim put a lot of work into this engine project and we are grateful that he is kind enough to share it with us.

A BIT OF HISTORY: James Atkinson, original designer of the engine we are about to build was born near Manchester, England in 1846.

Throughout his career he was involved with steam, gas and oil engines. He served his apprenticeship with a marine engineering company in Jarrow, a town near South Shields in England. Atkinson's interests were in steam engines and ships until about 1882. His interest in non-steam power first became evident in 1879 when he applied for a patent for a hot tube igniter. In 1881 and 1882 he filed patents for two stroke, Robson-type engines. In 1883 a company was formed to build these engines. The company was called, "The British Gas Engine and Engineering Co., Ltd.", and was located in London. Atkinson was appointed the managing director of this company.

To better understand what was going on during the period of time when Atkinson designed and built his "Cycle" engine we need to understand the challenge that he and other engineers of his time faced.

The popular and dominating engine of the day was the Otto four stroke engine. A man by the name of Nicolaus Otto had designed this engine and held patents protecting his invention. Otto's engine is often referred to as the father of all modern reciprocating engines. The Otto engine went into production at the opportune time of 1876. Customer demand soon exceeded productive capacity. These engines were used for hydraulic elevators, to power water pumps for water supplies, for railway service, to power factories and to produce power for electric lights. From the period of 1876 to 1890 Gasmotoren-Fabrick Duetz and its

Figure 2 The Atkinson "Cycle" engine made by Manlove, Alliott & Co., Ltd.; Great Britain. (From the collections of Henry Ford Museum &Greenfield Village).

licensee protected their position as the worlds sole supplier of Otto's engine. They ruled with an iron hand through litigation and threats challenging all builders who attempted to infringe on the Otto patents. The issues involved not only the four-stroke cycle, but any compression engine as well. Competition was virtually eliminated. Herein lies the challenge that Atkinson and other engineers of his time faced. They must develop another type of engine that could compete with the Otto four-stroke engine and at the same time not infringe on his patents.

ATKINSON INTRODUCES HIS "CYCLE" ENGINE: To meet the challenge of his era James Atkinson introduced his "Cycle" engine in 1886. What made it so unusual was that it did not have a separate cam shaft or timing gears, yet it was still capable of completing all four strokes in one revolution of the crankshaft. The engine had a single piston and included the desired feature of a longer and faster expansion on the power stroke. The result of this feature was that the Atkinson engine used even less fuel than the Otto engine. "The British Gas Engine Company" built over 1000 of these engines between 1886 and 1893. Manlove, Alliott and Co. in Nottingham built the "Cycle" engine between 1889 and 1892, as did the Warden Manufacturing Company in Philadelphia between 1889 and 1890. The original engines were painted dark green and the crank throw and governor were painted dark red. The engines were made in small sizes up to 30 horsepower. A one horsepower engine was supposed to have been made, but never placed on the market.

The invention of the incandescent light in the 1880's made it necessary to find a solution to a persistent problem. The newly developed lights flickered as the speed fluctuated between the firing impulses in the slower running gas engines. The Society of Arts wanted to find out which engine performed best in the area of power generation. In 1888 the Society conducted tests in London on several gas engines. These tests provided comparative data between the Atkinson "Cycle" and the Crossley-Otto four-stroke engines. One of the objectives was speed regulation and cyclic variation for electric power generation. During these tests the "Cycle" engine was shown to be superior and Atkinson was awarded the Society's gold medal.

Atkinson's fame was short lived however, because in 1890 the Otto four-stroke engine patent expired. This left the door wide open for many engine builders to switch to the four-stroke engine design. As competition became more intense the prices of the Otto type engines plummeted. This meant the finish for those manufacturers who chose to stick with a two-stroke design. They simply could not compete with the lower priced four-stroke engines. During this period only strong companies with proven products were able to survive.

Atkinson's company made one final attempt to introduce a two-stroke engine called the "Utilit'e" , but the venture failed. The British Gas Engine Co. finally went out of business in 1893.

This was not the end of Atkinson's career however. Crossley Brothers, Ltd. hired him in July of 1893 as chief engineer. He made several contributions at Crossley before retiring in 1912. He died in 1914, two years after his retirement from Crossley.

OPERATING PRINCIPLES: At first glance the absence of timing gears and separate camshaft may lead one to believe that the Atkinson "Cycle" engine is a two-cycle. This is not the case. Because of the unique two part linkage between the connecting rod and the crank throw, the piston is able to go through four complete strokes in one crankshaft revolution. The engine is water cooled and has a single piston that runs in a horizontal cylinder. The need for timing gears and separate cam shaft is eliminated by locating the cams on the crankshaft. The cams operate a pair of rather long and unusual push-rods. The push-rod action created by the cams operates the exhaust and intake valves located in the cylinder head. A spring loaded flyball governor controlled the fuel valve. A hot tube igniter was used for combustion. The intake to expansion ratio of Atkinson's engine was 0.56:1.

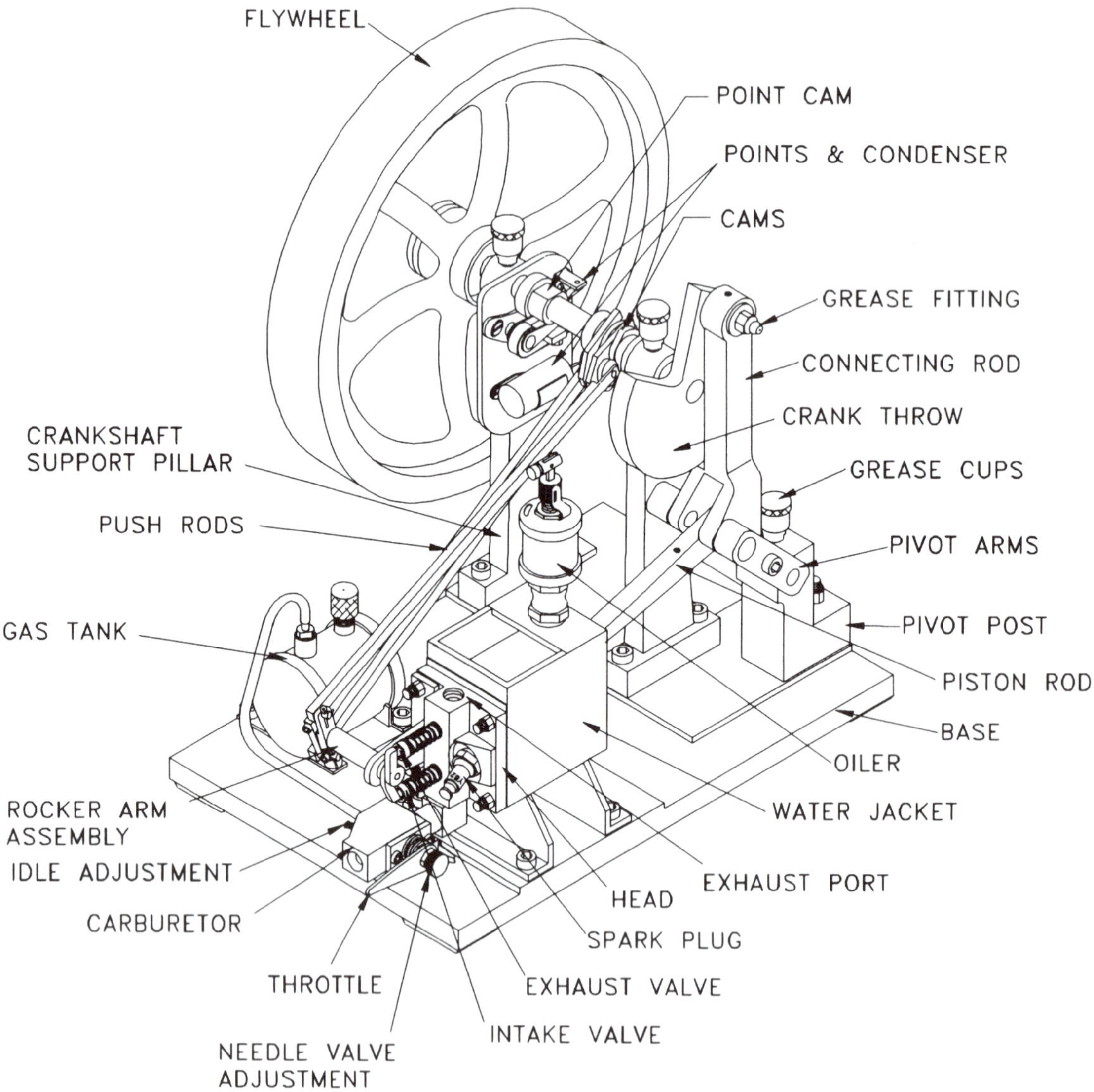

Figure 3

Before we begin discussing the operating principles of Jim's engine in more detail, have a look at the detailed drawing in figure 3. Figure 4 shows a photo of the same engine. You will notice that there are some cosmetic differences between the original and Jim's version and that Jim's uses spark ignition and has a carburetor. But the operating principles remain the same as those of the original Atkinson "Cycle" engine.

The flywheel is an aluminum casting and weighs about 3 pounds. Its finished diameter is 8-3/4". The 7/16" crankshaft is held in place by two cast aluminum support pillars. The ignition points and condenser are from a late 1970's Ford V-8 engine. They are mounted on the outside support pillar. An ignition cam mounted on the crankshaft opens and closes the points.The intake and exhaust cams are also mounted on the crankshaft. The lobes on the cams operate the pushrods and open and close the valves at the proper time. The crankshaft bushings are lubricated by grease cups.

Figure 4 A photo of the engine

The drive linkage consists of the crank throw, the connecting rod, the piston rod, the pivot post and the pivot arms. The crank throw is made from 3/8" H.R.S. and is brazed to the end of the crankshaft. The larger halfmoon end of the crankthrow helps balance the engine. The connecting rod is an aluminum casting and it pivots from the end of the crank. The piston rod is also an aluminum casting as is the pivot post. The pivot arms are made from aluminum bar stock. The pivot bushing is lubricated with a grease cup and the piston rod is lubricated with oil by means of oil holes at the piston connection and the crank connection.

The piston measures 1-1/4" in diameter and is made from 1-3/8" aluminum round stock. The piston rings used are 1-1/4" and are made from cast iron. The cylinder is made from 1-1/2" cast iron round stock.

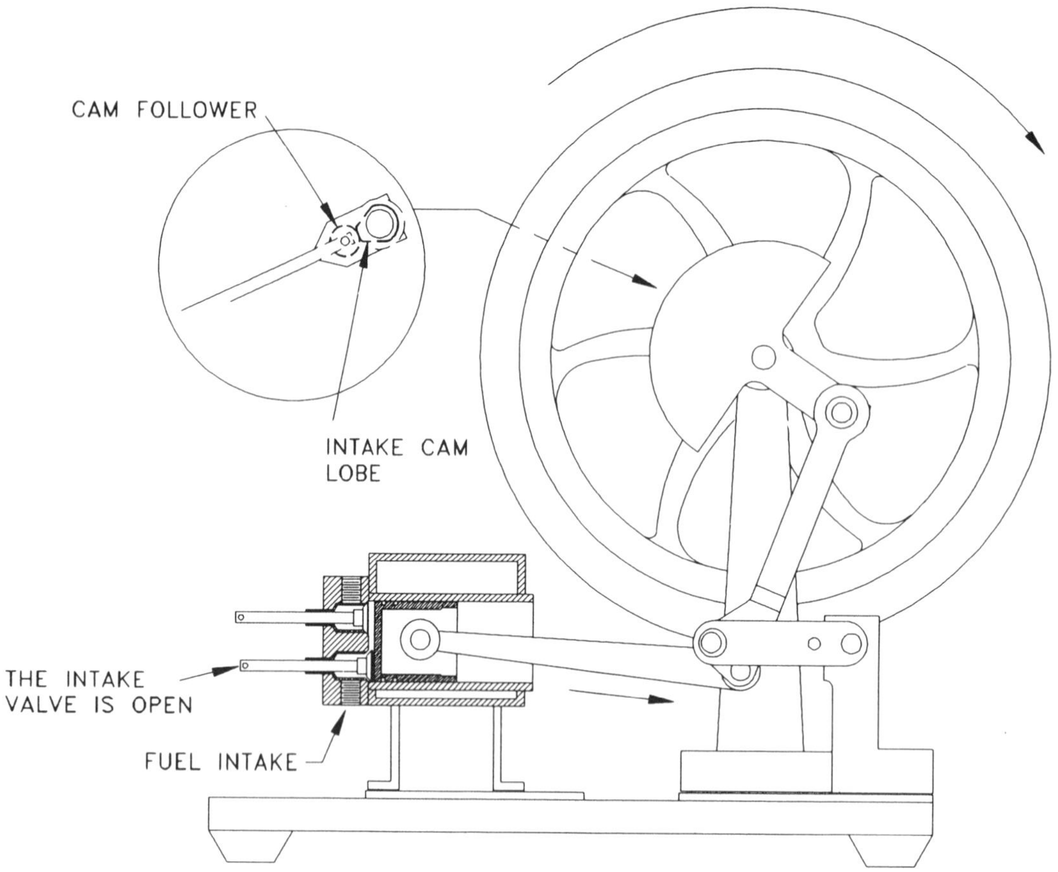

Figure 5

A water jacket is built around the cylinder and is constructed of 1/8" sheet steel. The water jacket is filled with antifreeze solution and helps keep the engine cool.

The cylinder head is an aluminum casting and is bolted to the end of the water jacket. It is drilled and tapped for a spark plug, valves, carburetor and exhaust.

The carburetor design is similar to those used on model airplanes and boats. The fuel flow is controlled independently by an adjustable needle valve. The air intake is controlled by an adjustable throttle barrel. The two are tied together so that when the throttle is opened the needle valve also opens slightly letting in more fuel for increased speed. The idle air mixture is also adjustable.

The gas tank is made from a short length of 2" diameter exhaust pipe and engine block freeze plugs are used for the ends. The inlet and outlet fittings are made of brass. The fuel used to power the engine is Naptha. Naptha is the same type of fuel used in some lanterns and camp stoves.

All of the components of the engine are mounted on a cast aluminum base measuring 10-5/8" x 7-1/8". As mentioned earlier, the engine is able to complete all four strokes in one revolution of the crankshaft.

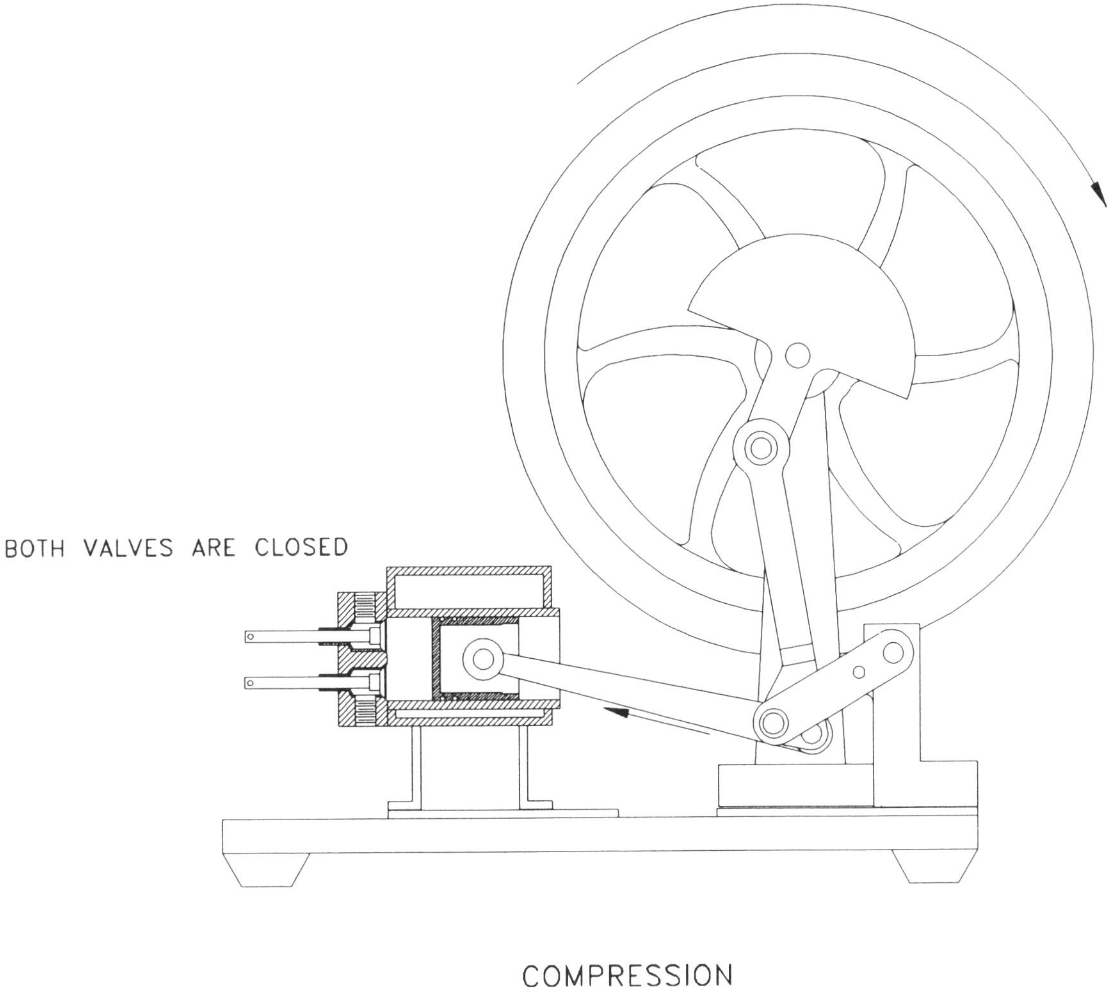

Figure 6

In order to better understand the 4-stroke sequence refer to figures 5 through 8. As we go through the sequence you will begin to appreciate how the crank linkage makes this engine so remarkable.

As the intake stroke begins, the lobe on the intake cam has just caused the intake valve to open and fuel is being pulled into the cylinder in vapor form. See fig. 5 for a visual of the intake stroke.

As the intake stroke is completed the compression stroke begins. Both valves are closed and the piston is moving forward and compressing the fuel mixture. See fig. 6 for a visual of the compression stroke.

At the end of the compression stroke the power stroke begins. The points break at top dead center and combustion occurs. See figure 7 for a visual of the power stroke.

At the end of the power stroke the exhaust stroke begins. The lobe on the exhaust cam forces the exhaust valve open and the burnt fuel mixture is forced out through the exhaust port. The distance between the top of the piston and the head at the top of the exhaust stroke will be .020". See figure 8 for a visual of the exhaust stroke.

The linkage action is fascinating to watch and when you build your engine you will be amazed at how sensitive the location of each linkage component is. For example, moving the pivot post as little as 1/32" in any direction will dramatically affect the linkage travel pattern. Slight differences in hole locations in the various linkage parts will also dramatically affect travel patterns.

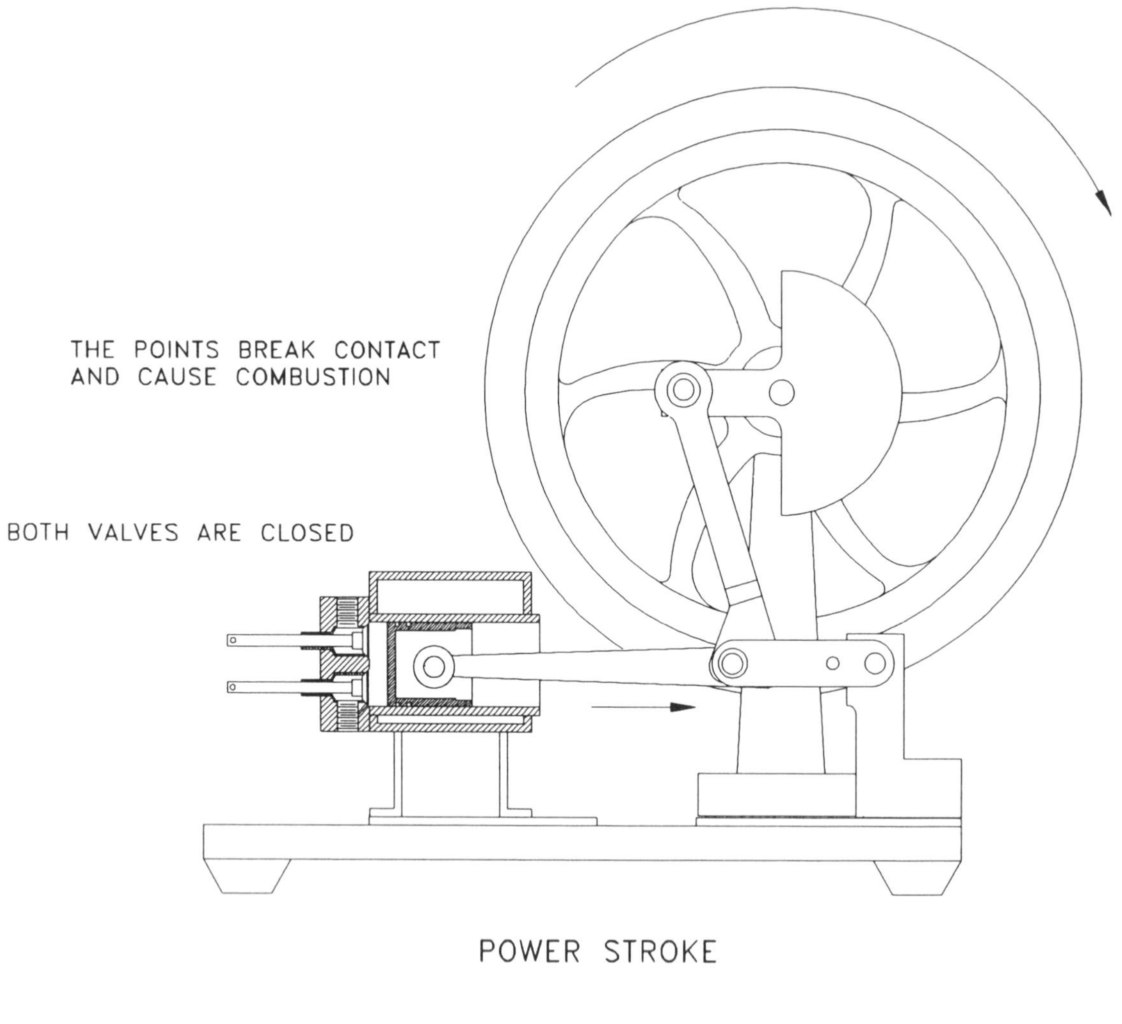

Figure 7

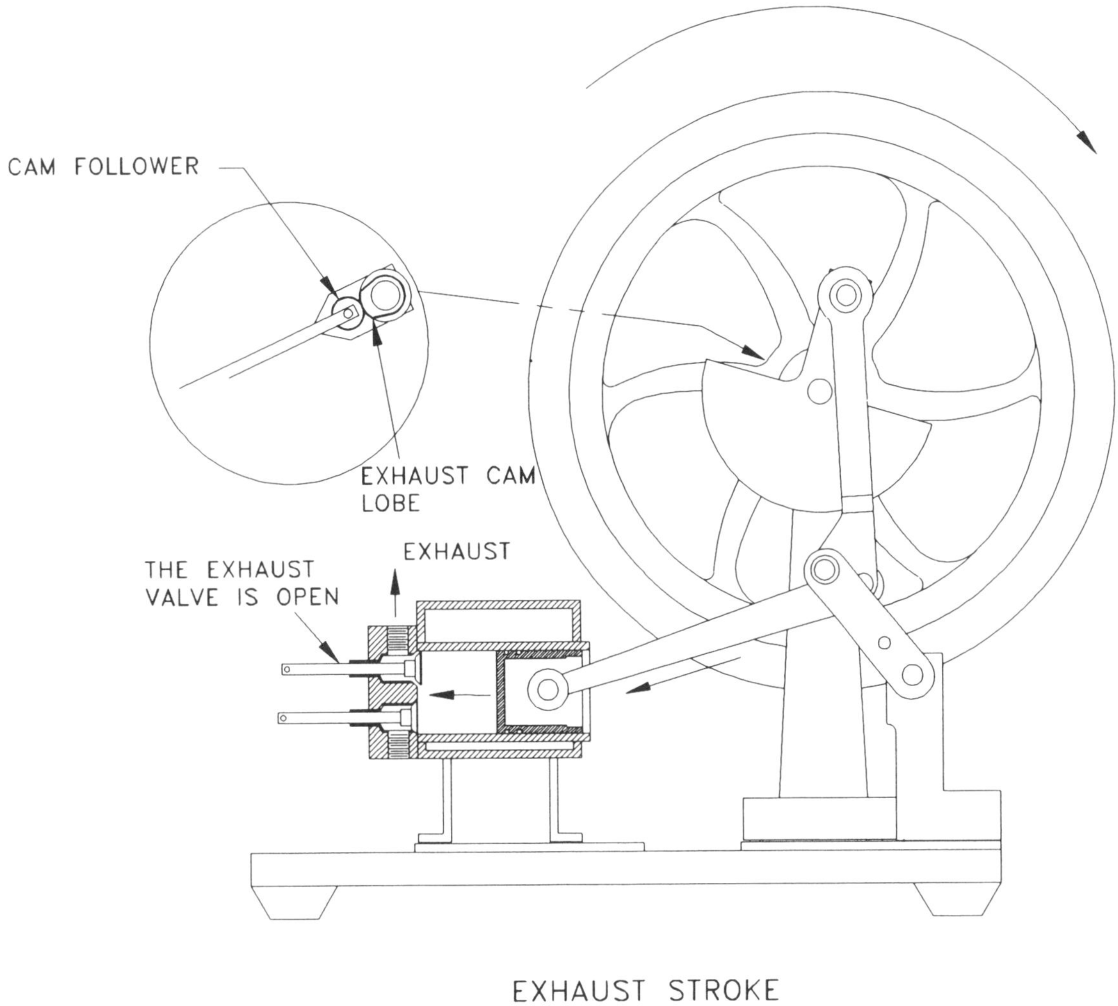

Figure 8

PREPARING TO BUILD: Because aluminum castings are suggested in this project, the flywheel, base, the two support pillars, the pivot post, the connecting rod and the cylinder head are going to be the main consideration before beginning.

We have included the details for producing patterns and molds for those of you who have a foundry setup in the shop. You will find that these patterns are simple to make and that the molds will be easy to form and pour. But those of you who do not have a foundry, and can't be talked into putting one in have an opportunity to choose from a couple of other alternatives. First of all, none of the parts are so complex that they could not be made from stock material. The base could be made from a 1/2" thick slab of aluminum or steel faced off on a milling machine. The other parts could be made from weldments or formed from blank stock with the help of a lathe and milling machine. Probably the more challenging would be the flywheel. And there are alternatives here too. Jim Lewis modified a flywheel taken from an

Figure 9 Side View

old lawn mower and used it on his engine. Some have made flywheels by welding spokes and a hub in the center of a section of pipe and then turning the outside diameter on a lathe. Some companies such as Power Model Supply, 13260 Summit Drive, NW Cor Hwy 67 & Long Rd., Desoto, MO 63020, Phone (314) 586-6466 offer a selection of cast flywheels for sale. Power Model Supply also sells the 1-1/4" piston rings needed for this project.

A sturdy workbench and vise will be used throughout the project as well as an assortment of handtools including a hacksaw, files, scriber, square, rule, calipers, drills, reamers, taps, dies, screwdrivers, allen wrenches, wrenches, "C" clamps, etc. Some basic woodworking tools will also be required if you make the molding patterns.

A metal lathe is used to make many of

Figure 10 End View

the parts. We used a lathe with a 9" swing to make the flywheel, but many of the other parts were made on a 6" lathe. Normal lathe accessories are required such as a 3 and 4-jaw chuck. A tail stock chuck will be needed for many of the drilling and threading operations. A tail stock die holder will also be needed for some of the threading operations.

Except for milling the surface of the base, all of the other milling operations can be performed using a milling attachment mounted on a lathe.

Other equipment required will be a welder capable of 75 amps and a drill press. Some brazing and soldering is also necessary so a torch will be required.

Several jigs and fixtures will be needed, but we can make those ourselves and we will cover that in a separate section.

SAFETY: It is impossible to point out all of the dangers that you are likely to encounter while working on this project. The most important thing is to take your time and think each move through carefully. Pause before each step and ask yourself, is there any way that this tool or machine can hurt me and then take the proper precautions.

One of the most common injuries encountered are foreign objects in the eyes and mangled fingers and hands. This should tell you that it is important to wear eye protection and to consider revolving drills, saw blades, mill cutters, flywheels and the like to be your worst enemy. Remember that a small wood or metal chip in the eye can cause permanent blindness and that only a moment of carelessness around a machine can cause the loss of a finger or hand or worse. Also remember that long hair, jewelry and loose fitting clothing can get caught in a machine and cause serious injury and even death. there are many other dangers that will be realized and avoided by using common sense.

Understand all of your machinery and make sure it is in good working order. Keep the floor swept clean and post clear warnings as reminders to yourself and others.

SOURCES FOR MATERIAL: Every time we begin a project it seems that material and supplies get harder and harder to find at the local level. Over the years I have really learned to appreciate our Mail Order Suppliers. Without them a lot of these projects would not be possible. We have found the best way to find and keep up to date on sources for materials is to subscribe to "The Home Shop Machinist" magazine. They regularly carry ads from all the best sources. Write them at PO Box 1810, Traverse City, MI 49685 for subscription information.

MATERIAL LIST:

1- Flywheel casting

1- Base casting

1- Crank casting

1- Pivot post casting

1- Cylinder head casting

2- Crankshaft support pillar castings

1- Piston rod casting

1- Piece of 3/8" x 4" x 4" mild steel flat bar. (To be used for the crank throw)

MATERIAL LIST CONT:

1- Piece of 1/8" x 1' x 1' sheet steel. (To be used for the water jacket)

1- Piece of 1/8" x 1-1/4" angle 7" long. (To be used to make legs for the water jacket)

1- Piece of 3/8" x 3/4" x 5" aluminum flat bar. (Used to make the control links)

1- Piece of 1-3/8" diameter aluminum round rod 6" long. (To be used to make the piston)

1- Piece of 1-1/2" x 6" cast iron round rod. (To be used to make the cylinder)

1- Piece of 1/16" x 6" x 6" sheet steel. (for rocker arm bracket and mounting plate for the points)

1- Piece of 1-1/2" x 4" C.R.S. Round rod. (For the tapered flywheel hub insert)

1- Piece of 1" x 3/4" aluminum bar 2-1/4" long. (Used to make the carburetor)

1- Piece of 3/4" brass round rod 6" long. (For bushings and other small parts)

1- Piece of 5/8" x 6" brass round rod. (For bushings and other small parts)

1- Piece of 1/2" x 6" brass round rod. (For bushings and other small parts)

1- Piece of 3/8" x 6" brass round rod.

A 1' length of each of following. 1/4", 3/8", 5/16", 7/16", 1/2", 5/8" and 1" water hardened round drill rod.

1- Piece of 1/8" drill rod 6" long. (For valve stems)

2- Pieces of 1/16" x 1/4" long round rod. (For valve spring keeper pins)

2- Pieces of 3/16" x 1' key stock. (For pushrods)

1- 3" piece of 1/2" aluminum round rod.

1- 6" piece of 5/8" aluminum round rod.

1- 6" long piece of 1/8" brass tubing. (for gas tank outlet)

1- 2" long piece of 1-3/4" ID x 1-5/8" OD exhaust pipe. (For the main body of the gas tank)

HARDWARE:

2- Springs .312" O.D. x .286" WD x 1" long. (valve springs)

1- 1/4-20 x 1-1/2" bolt w/nut & flat washer.

1- 5/16-18 x 1" bolt.

4- #10-24 x 1" socket head cap screws.

2- #8-32 x 1/2" C.S. screws.

1- 7/16" friction washer.

3- #6-32 x 3/8" socket head set screws.

1- #10-24 x 1-3/4" socket head cap screw.

1- 1/4" flat washer.

1- 1/4-28 grease fitting.

HARDWARE LIST CONT:

4- #10-24 x 3/8" socket head cap screws.

4- #10-24 x 3/4" studs threaded both ends.

3- #10-24 x 3/4" socket head cap screws.

3- #8-32 x 1/2" screws.

1- 2" x 2" piece of paper gasket material. (for head gasket)

2- 1/16" x 1/2" cotter pins.

1- 1" x 1/8" drip oiler.

1- Set of ignition points and condenser from a late 1970's - early 1980's Ford V-8.

1- Champion #Y82 spark plug or could also be an NGK CM-6 spark plug.

1- Ignition coil. (Could be auto or lawnmower)

1- Small 6 volt rechargeable battery.

1- #4-40 x 7/16" Phillips head screw.

1- Piece of 1/4" brass hex stock 2" long.

1 #8 32 x 1" brass screw.

1- #8-32 brass knurled nut.

1- #18 darning needle 2-3/8" long.

2- 5/16-24 nuts 3/32" thick.

1- 1/8-27 lock nut 3/32" thick.

1- 3/8" long x .5" OD x .458" ID spring.

2- 1/4-28 lock nuts 3/32" thick.

2- #4-40 x 1/2 screws.

2- #P38 freeze plugs measuring 1-5/8" ID x 1-7/8" OD. (These are the ends for the gas tank)

PRODUCING THE CASTINGS: We must assume that you are familiar with the foundry operation and have the equipment and tools needed to produce the patterns and castings for this project.

For those of you who are not familiar with the pattern making and casting process, a brief description follows. If you decide you might be interested in adding a foundry to your shop and want to find out more, check out some of the great books on the subject available from the suppliers listed in the back of this book.

The castings in this project are produced using sand molds. The procedure is to embed a pattern in sand in a two part mold. When the pattern is removed from the sand it leaves an impression in the mold. It is necessary to open the mold, remove the pattern and then close the mold before the molten metal can be poured. Because of this the molding sand must be able to hold its shape and bond together. To gain this characteristic the sand is mixed with an adhesive agent. In the case of *GREEN-SAND* molding the adhesive agent may be clay and water. Another type of molding sand available is a petroleum based oil bounding sand. Either of the above types of molding sands will work fine for this project.

The molds in this project are made in two part flasks. The flasks are really just wooden boxes with no bottoms or tops. The flasks fit together with pins at each end so

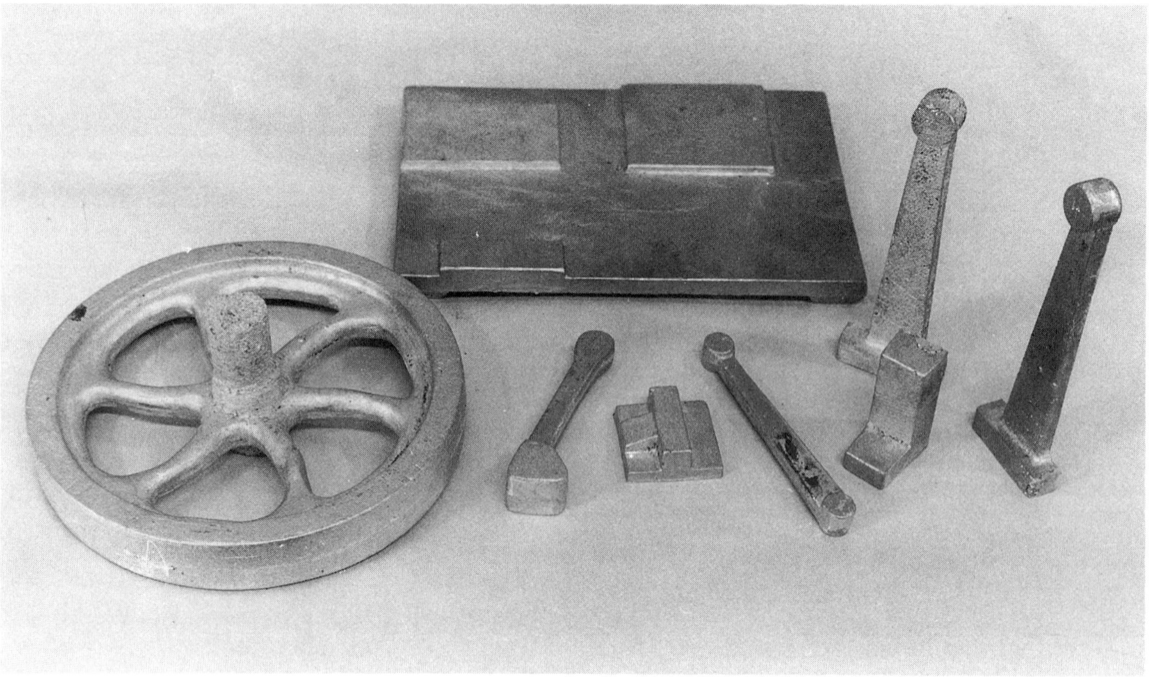

Figure 11 The castings

that the halves can be aligned in perfect register each time they are put together. The bottom half of the flask is called the drag and the top half of the flask is called the cope. See figure 12 for a drawing of a typical flask.

In brief, a typical molding process consists of placing a pattern on a flat board called the molding board. The drag half of the flask is inverted over the pattern. The pattern is coated with a small amount of parting dust to prevent the sand from sticking to it. The flask is then rammed full of molding sand and struck off level. A bottom board is rubbed in over the sand. Molds must be vented to release steam and gas. This is done with a piece of 1/16" wire inserted into the mold and almost to the pattern. After venting, the drag is rolled over and the molding board is set aside. The bedded pattern is now visible. The cope half of the flask is set in place on top of the drag. Another light dusting of parting dust and then a sprue pin is set in place. The purpose of the sprue pin is to open the way for molten metal to enter the cavity left in the mold when the pattern is removed. The cope is then rammed full of sand and struck off level. After the sprue pin is removed the cope is lifted straight up and set on edge behind the drag. All loose dust is blown away and the edges of the mold around the pattern are strengthened by swabbing with soft bristle brush dipped in water. After cleanup the pattern is rapped lightly and removed from the mold. A screw eye screwed into the center of the pattern aids in its removal. The mold is then reclosed and poured.

There are some variations to the above procedure depending on what one is casting. For instance, some of the patterns in

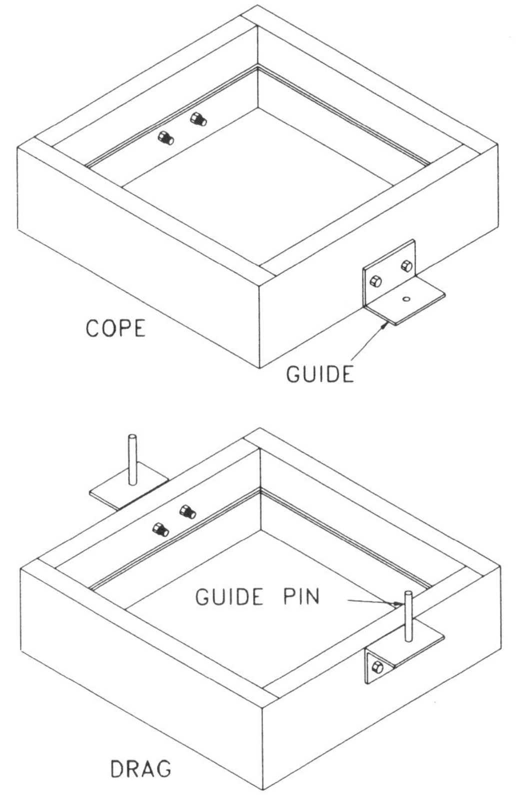

Figure 12 Typical homemade flask

this project are split. This simply means that the pattern is made in two halves split along the natural parting line. The two halves are held in perfect register by pins called pattern dowels. The pattern dowels are installed so the pattern can be put together in only one way. The molding procedure would be to ram up one half of the pattern in the drag. After the drag is rolled over the pattern halves are put together and the cope is rammed up as before. When the mold is separated there will be a pattern half in both the cope and the drag.

MAKING THE PATTERNS: These patterns all easy to make and only basic woodworking skills will be required. Because white pine is common and inexpensive in our area we used it to make most of our patterns. When necessary we use carpenters glue to join pattern parts. Polyester auto body putty can be used for filling imperfections and wiping fillets on inside corners of the pattern. After the patterns are complete they must be sanded smooth and painted or varnished to prevent moisture absorption. The base, head and crankshaft support pillars are single patterns and the flywheel, connecting rod, pivot post and piston rod are all two part patterns split at the natural parting line.

In order to remove the pattern from the mold without damaging the sand impression it is important that the vertical edges of the pattern have sloped edges. This is referred to as giving the pattern draft. A minimum draft of 1 degree is fine for a skilled molder. But a beginner may require as much as 3-5 degrees in order to insure a clean cavity. You will notice as you proceed that the draft direction is noted on each pattern drawing.

When making the patterns, outside edges must be given a radius and inside corners given a fillet. The object is to eliminate any sharp edges or angles that could weaken the mold.

Molten metal shrinks upon solidification and castings are often machined to their final size. We have made allowances in our pattern dimensions to compensate. Aluminum usually shrinks approximately 3/16" per foot in all directions. That means a pattern that is 12-3/16" long will produce a casting 12" long.

The most important thing to remember when making these patterns is to allow plenty of draft and make sure that their surface is smooth.

THE FLYWHEEL: The flywheel with its curved spokes certainly adds to the appearance and is fitting to the style of the Atkinson engine.

The two piece split pattern is easy to make and is shown in figures 13 through 15. The two halves are simple flat back patterns and are split at the natural parting. A pair of pegs align the patterns in perfect register.

The original pattern was made from 5/8" thick, C grade, exterior fir plywood and has held up for several years now and has produced dozens of molds and is still in excellent condition. Fir plywood is not the best material for making patterns, but by filling the imperfections with auto body putty, doing a little sanding and then finishing with a couple of coats of varnish you can create an acceptable pattern.

The pattern layout is shown in figure 13. The method used to do the layout is a simple sequence done with a compass. Small washers can be used to mark the fillets. The pattern consists of two 9" diameter plywood disks. It is best to mark the spoke pattern on one of the halves and then clamp the two together to cut out the spoke openings. By doing it this way you will have only one layout and are assured a matched pair.

First, divide the 9" circle into 6 equal parts. This can be done with a protractor or a 60 degree triangle. Use a compass to mark the 7-3/4" circle representing the inside radius of the outer rim. Next mark the 6-3/4" circle representing the inside rim. Referring to step 2 in figure 14, set the compass at 6-3/4" which is the diameter of the inner circle. Pivot the compass at point A and draw arc AB. Continue around the circle drawing an arc at each division line.

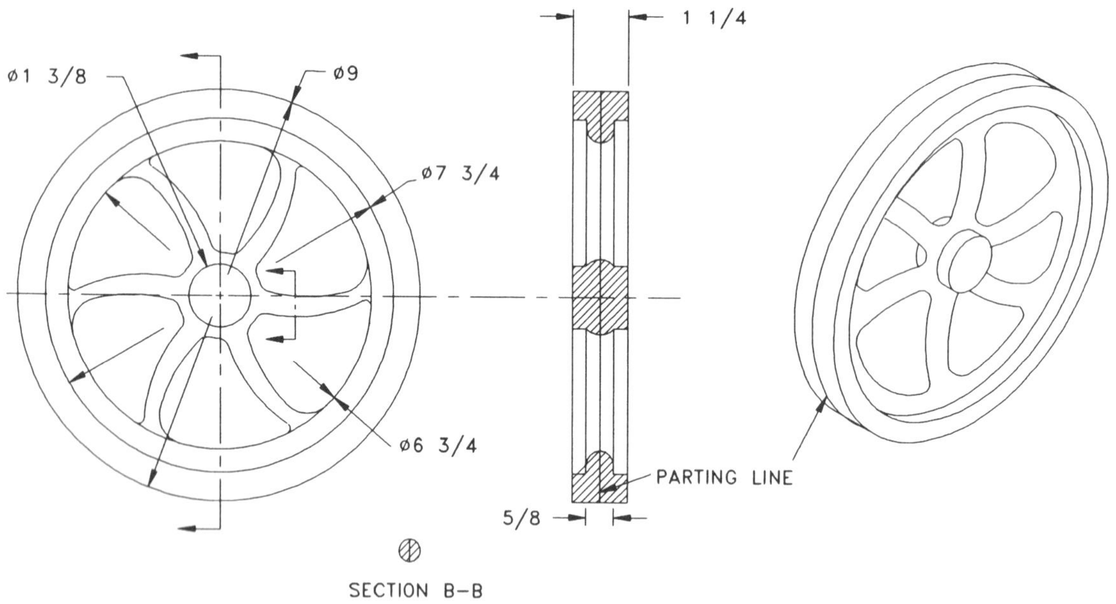

Figure 13 The flywheel pattern

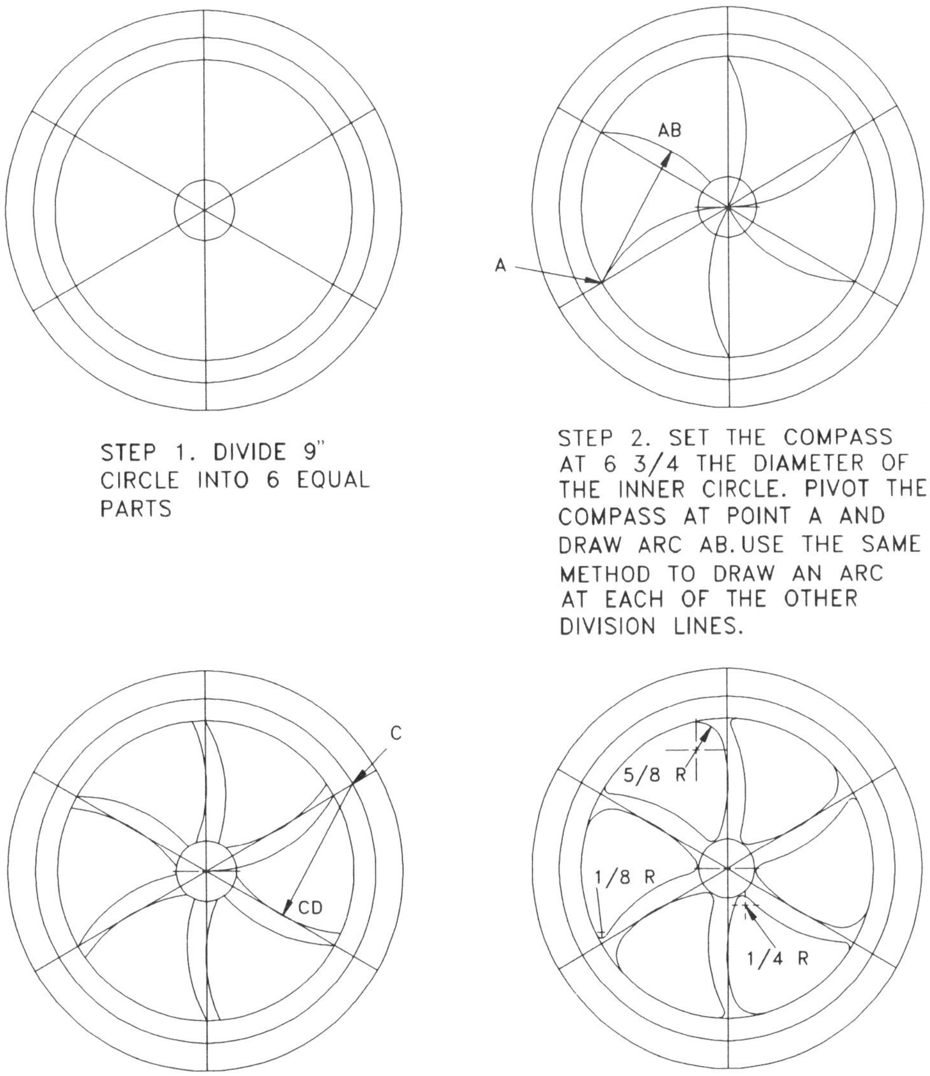

STEP 1. DIVIDE 9" CIRCLE INTO 6 EQUAL PARTS

STEP 2. SET THE COMPASS AT 6 3/4 THE DIAMETER OF THE INNER CIRCLE. PIVOT THE COMPASS AT POINT A AND DRAW ARC AB. USE THE SAME METHOD TO DRAW AN ARC AT EACH OF THE OTHER DIVISION LINES.

STEP 3. WITH THE COMPASS STILL SET AT 6 3/4 PIVOT FROM POINT C AND DRAW ARC CD. CONTINUE AROUND THE CIRCLE DRAWING ARCS AT EACH DIVISION LINE.

STEP 4. DRAW ARCS TANGENT WITH THE SPOKES AND RIM AND WITH THE SPOKES AND HUB

Figure 14 **Laying out the flywheel pattern**

Referring to step 3 in figure 14 and with the compass still set a 6-3/4" pivot from point C and draw arc CD. Continue around the circle drawing arcs at each division line. Finally draw the fillets tangent with the spokes and rim and spokes and hub. As mentioned earlier these are not critical and can be drawn with washers or other round objects close to the sizes indicated.

With the layout complete, clamp both pattern halves together. Mark them in such a way that they can be reassembled in the exact position later. Drill starting holes and cut the spoke openings with a jig saw.

Mount either pattern half on a lathe face plate and turn the hub and rim detail.

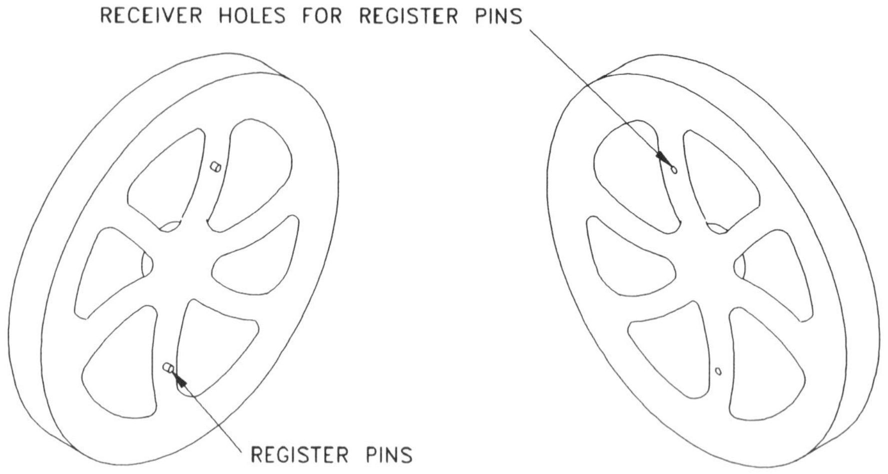

Figure 15 The two halves of the flywheel pattern

Center the blank carefully in relation to the original layout center so the pattern will be truly symmetrical. Remember to draft the pattern at each diameter. Make the opposing second half in the same manner.

Now round the spokes and the inside rim as shown in section BB in figure 13. This can be done by hand, or if you have a hand grinder that can be fit with a small sanding drum you can use it to speed the process up a little. Finally clamp the halves together in their original position and drill through for the alignment pegs. The alignment pegs can be made from a nail or piece of 1/8" round rod. Insert them in the holes and secure them with epoxy. The location of the alignment pegs is not critical. See figure 15 for an approximate position. Fill any flaws, sand the pattern smooth and seal with two or more coats of varnish.

Lay the pattern half without pegs face down on bottom board. Sprinkle with parting dust and ram the drag up over it. Rub in a bottom board, vent the mold and roll the drag over. Lay the second half of the pattern over the first making sure that the pegs line up with their mating holes. A good place to gate the mold is at the hub. We used a 1-1/4 round dowel for a sprue. Place a register pin in the bottom of the dowel and drill a matching hole in the center of the hub. This will register the dowel with the hub and help keep it in position as you ram up the cope.

Vent the mold, remove the sprue and rub in a bottom board so the cope can be laid down flat to rap and draw the pattern half. Remove both pattern halves from the cope and drag. They should come out easily requiring very little clean up. Reclose the mold and pour. Since this is a large casting, there is a need to weight the cope to prevent molten metal from escaping at the parting line.

THE BASE PATTERN: The base pattern is shown in figure 16. It is made from 1/2" plywood. The raised pads are made from 1/4" plywood. The pads are attached with wood glue and screws. The screws are countersunk and the holes are filled in with body putty. The legs are made from white pine and are also attached with glue and screws. After the pattern is made and the glue has dried, fill all imperfections with body putty and make sure to fillet all of the inside corners. Sand smooth and coat with

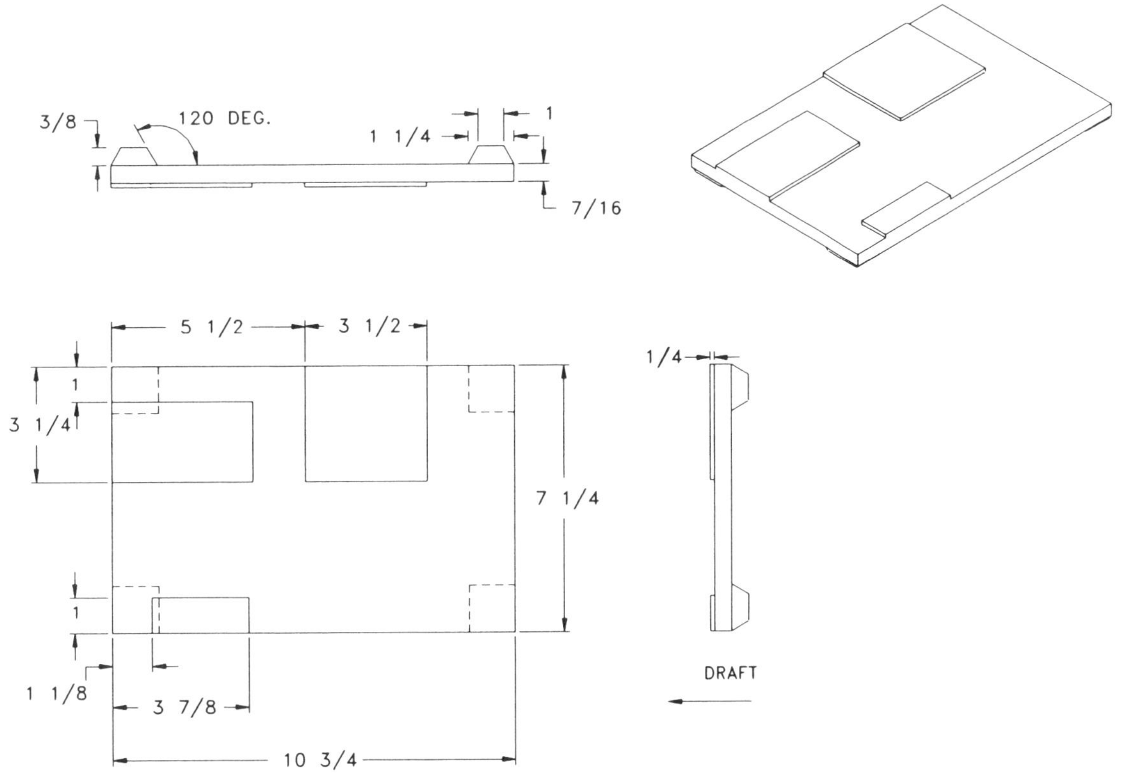

Figure 16 The base pattern

two or more coats of varnish.

The molding procedure calls for a follow board which is simply a bottom board with holes cut out for the legs of the base so the pattern can rest flat on the bottom board while the drag half of the mold is being rammed up. Set the base on the follow board and set the drag half of the flask in place. Sprinkle with parting dust and ram up the drag. Vent the mold, rub in a bottom board and roll it over. Remove the follow board, set a sprue pin at one end of the base and ram up the cope. Vent the cope, remove the sprue, separate the flask and remove the pattern, clean up the mold, cut the gate, Reclose and pour.

THE CONNECTING ROD PATTERN: The connecting rod pattern is shown in figure 17. It is a two part split pattern and is made from white pine. Mark the side profile on one of blanks, clamp the two blanks together and cut to size. Drill through for the register pins and epoxy them in place. Mark the back profile and cut to size. Fill all imperfections with body putty and round all corners to radius shown. Sand the pattern smooth and finish with two or more coats of varnish.

Set the pattern half without pegs face down on a bottom board and ram up the drag. Vent and roll the drag over and place the second half of pattern on the first

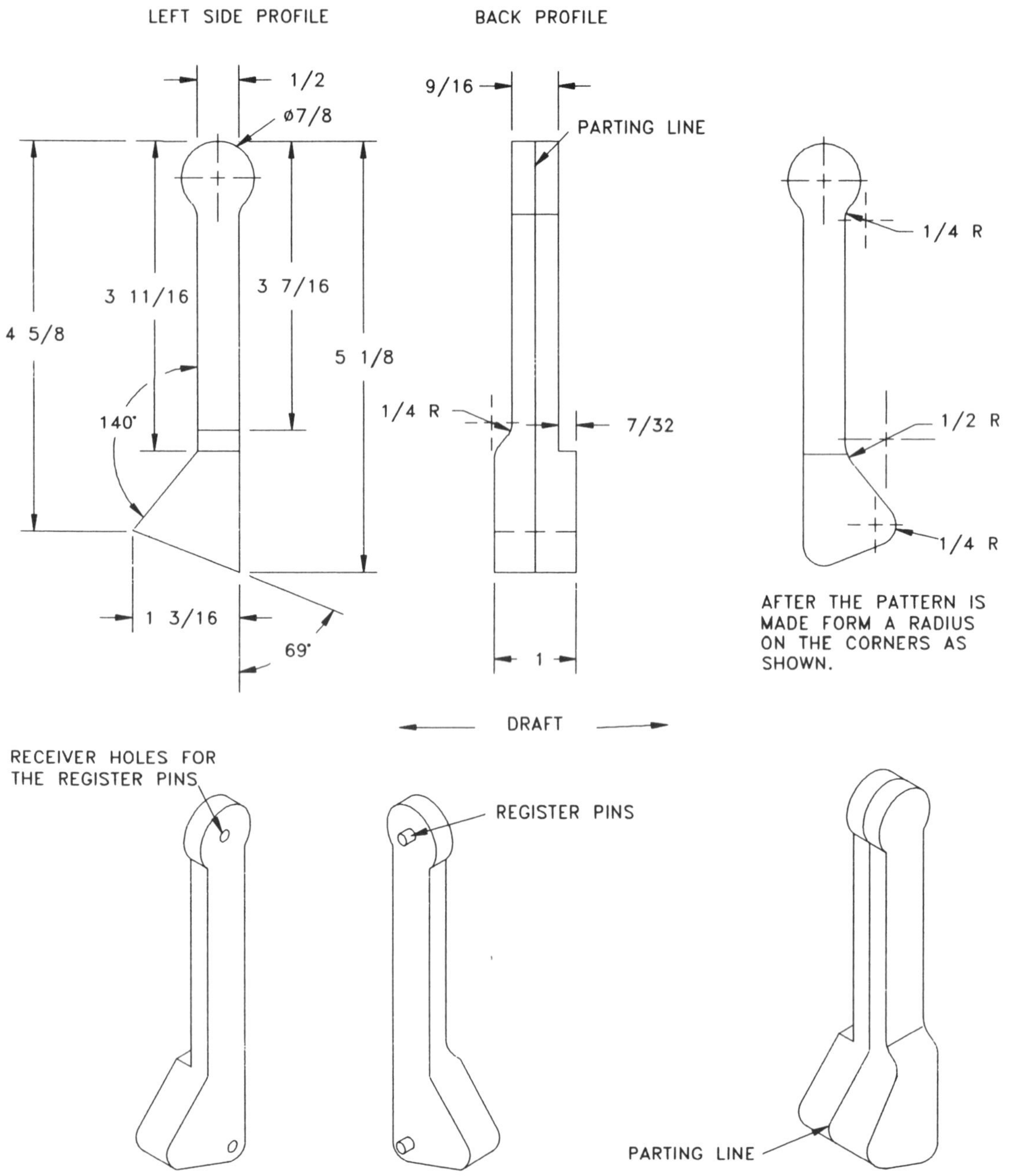

Figure 17 The connecting rod pattern

making sure the pegs are aligned properly. Position the sprue and ram up the cope. Remove the sprue and separate the cope and drag to remove the pattern halves. Clean up the mold, Remember to cut the gate, r- lose and pour.

THE PIVOT POST PATTERN: The pivot post pattern is also a split pattern. It is shown is figure 18. It is made from white pine and is sanded smooth and finished with varnish. The casting procedure for it is the same as the connecting rod.

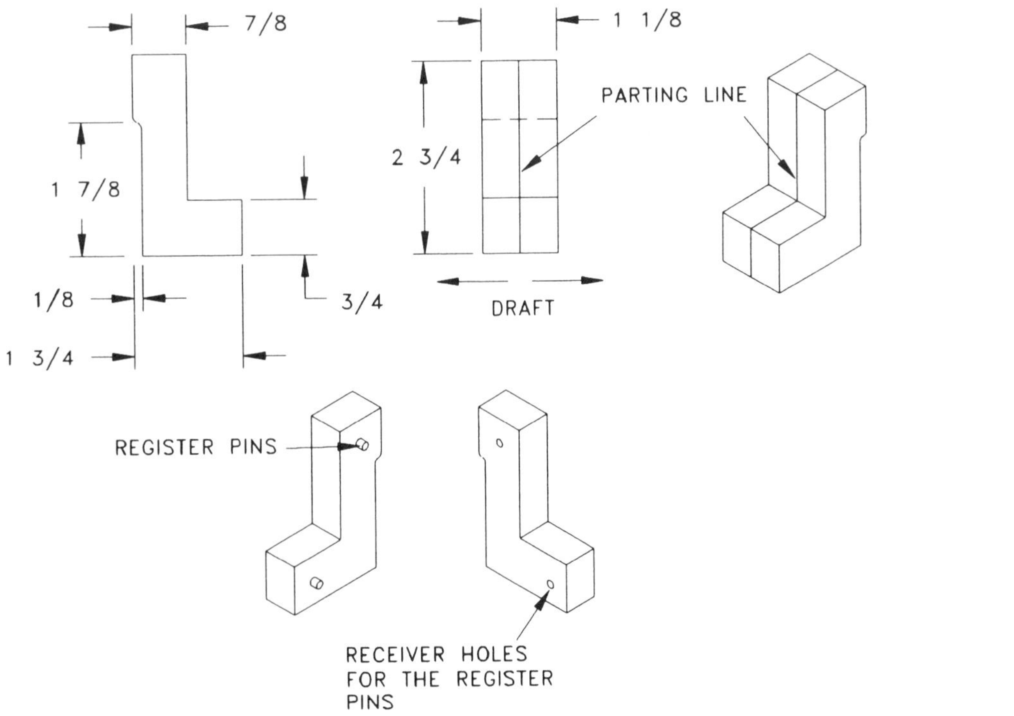

Figure 18 The pivot post pattern

THE HEAD PATTERN: The head pattern is a simple one part pattern made of white pine. It is shown in figure 19. The angled portion of the pattern can be made separately and glued in place. Be sure and fillet all corners. Sand smooth and finish with varnish.

Lay the head flat on the bottom board and ram up the drag. Rub in a bottom board, vent and roll over. Set the sprue and ram up the cope. Remove the sprue, separate the flask and remove the pattern. Clean the mold cavity, cut

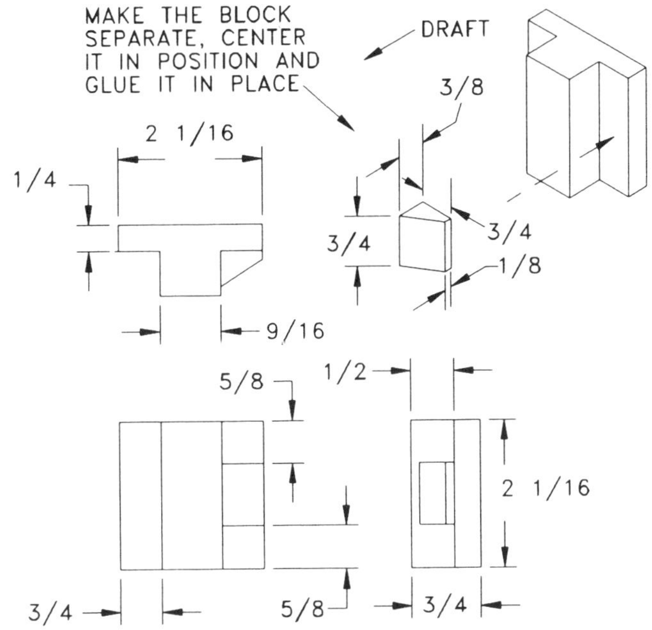

Figure 19 The head pattern

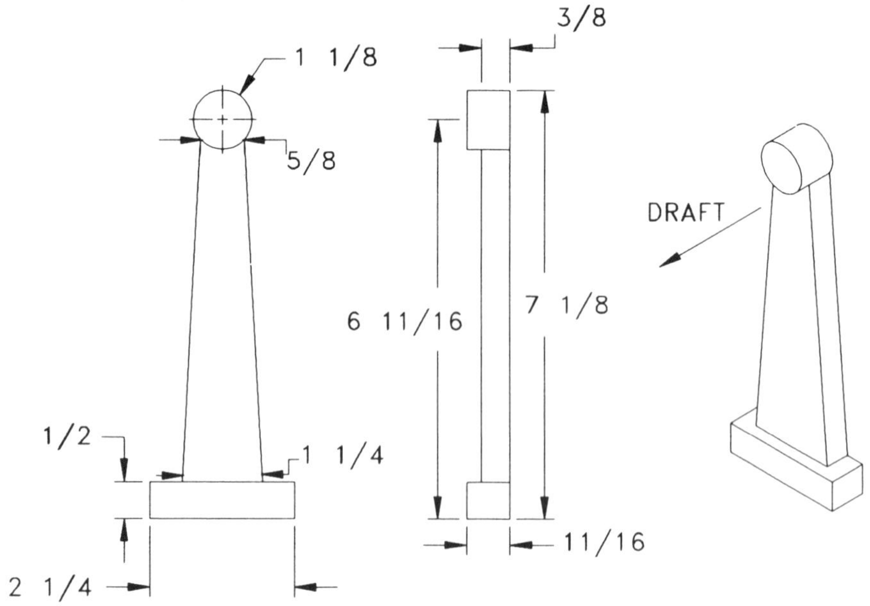

Figure 20 The crankshaft support pillar pattern

the gate, reassemble the flask and pour the casting.

THE CRANKSHAFT SUPPORT PILLAR PATTERN: The pattern for this casting is also a one part pattern. The center post section is cut from white pine. The 1-1/8" diameter round boss is cut from a piece of wooden dowel and glued in place. The base is also made separately and glued in place. Fillet all of the inside corners. Sand smooth and varnish. Prepare the mold and pour the casting. Two pillar castings are required. See figure 20 for pattern dimensions.

THE PISTON ROD PATTERN: This is another split pattern. It's simple to make. The ends are formed by gluing a 5/8" diameter disk that is 1/8" thick on the side of each end. Figure 21 shows a drawing of the pattern. Sand and varnish the pattern and prepare the mold and pour the casting.

When the castings have cooled, cut away the sprue and use a file to clean up the rough edges. A 16" double cut file is ideal for aluminum. Figure 11 shows a photo of all of the castings and how they look after cleanup.

CONSIDER MAKING THIS A GROUP PROJECT: Before beginning the construction of the engine it is worthwhile considering the possibility of making this a group project. Two or more heads are better than one. For instance, one person may have a foundry setup. Another can make the patterns and yet another may have a milling machine. Then consider how much can be gained by pooling and sharing knowledge. There are many local metal working groups across the country and more forming daily. With a little investigation you may find such a group in your area. And if not, consider forming one. Often all it takes is a small ad in the local paper.

Another phenomenon that is rapidly

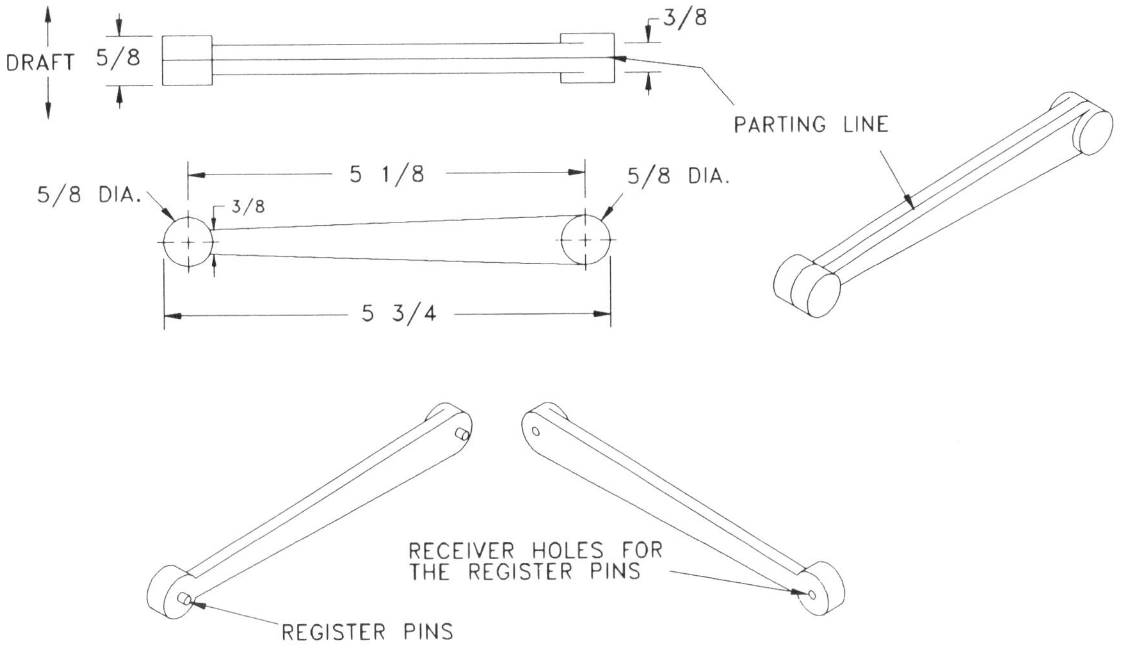

Figure 21 The piston rod pattern

taking hold is the internet. Through this medium it is possible to share information and put together group projects with people all over the world.

At the time of this writing there is one such project going on in the Handcraft/MetalShop forum on Compuserve. It's a group of about 20 guys from around the country and they are building a small dividing head. One guy in California made the pattern and then shipped it to another guy in Missouri who then poured the castings for the other members. Of course everyone shares in the cost and they all keep each other informed and inspired. You can imagine how much fun this can be and how it can make impossible projects suddenly seem possible.

MACHINING THE PARTS AND ASSEMBLING THE ENGINE: Not all of the parts in this project require machining, but a good many do. As in any engine project certain parts must be made to close tolerances. We found this project to be very forgiving, however certain areas such as carburetor construction, piston and piston ring fit, seating the valves and the crank linkage alignment to be critical. This is not to say that these are difficult items to produce it just means that special care is needed to produce accurate parts.

SPECIAL TOOLING: On an engine project such as this quite a bit of special tooling is needed most of which we can make right in the shop. The tooling will include such things as threaded arbors, collets, a cylinder lap, special tooling to make the valve cams and more.

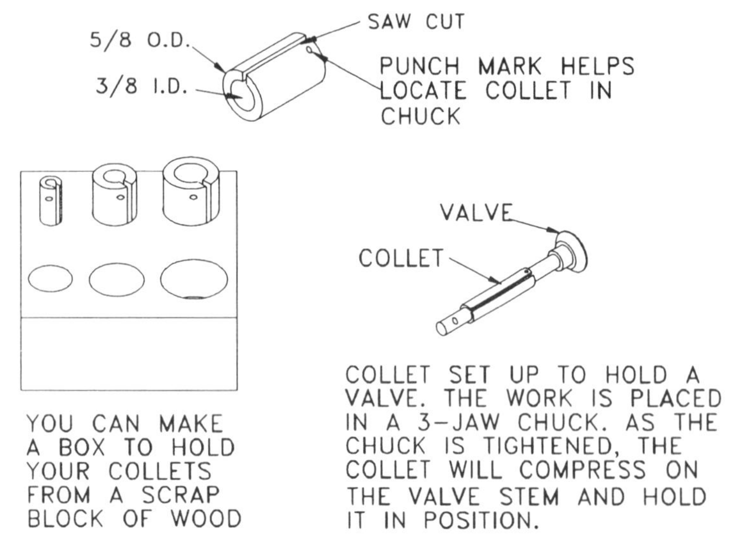

Figure 22 Shop made collets

One of the accessories you will need are collets and we can discuss those right now. Collets are clamps that are used to securely hold round rod in the lathe. They have the advantage of always holding work on center which saves a lot of set up time. They are especially handy for holding small work. A chuck and set of collets is an expensive investment to say the least. If you already own a set you're in luck, but if you don't you will really appreciate the savings gained through making your own. Whether what follows is an original idea or not, I don't know. It's just another one of those great tricks my Dad came up with, and it sure was a lifesaver in this project.

Figure 22 shows a drawing of a shop made collet and an example of one in use. The one in the drawing is made to hold 3/8" stock. The outside diameter is 5/8" which leaves an outside wall thickness of 1/8". If the outer wall is any thicker the chuck will not grip tight enough. The smallest wall thickness acceptable is 1/16" so this blank could have been used to make a 1/2" collet. After the collet has been drilled and reamed to size it is sawed through on one side at a marked reference point so that it can be placed in the chuck in the same position every time. The saw cut gives the collet room to compress around the work as the chuck is tightened. These collets have worked very well for us and hold the work tight and as long as they are set up in the chuck in the right position they will turn true every time. There is one disadvantage to these collets and that is that they only work on the lathe and chuck they are made on.

Collets can be made from mild steel or aluminum round rod. To make one, first cut off a length of blank stock 1" long. Place the blank in a 3-jaw chuck to face off both ends. Drill and ream to desired size. Before removing the collet from the chuck put a reference mark on it that corresponds with a mark on the chuck. You can use a center punch to make the mark. The collet will be sawed through at this mark. When you place the collet back in the chuck all you have to do is align the cut with the reference mark on the chuck and the collet will align on center. It will also be important to know which end of the collet goes in the chuck first, so locate a punch mark at the front of the collet and to one side of the cut line.

It won't be long before you accumulate a large number of collets. You can make a holder for them as shown in figure 22 out of a scrap piece of lumber.

I will show you how to make the rest of the tooling required as it is needed.

THE CRANKSHAFT SUPPORT PILLARS: The pillars are shown in figure 23. The first operation is to drill and ream the 9/16" hole in each one for the crankshaft pillars. It is important that these holes be aligned so put the two pillars together back to back so they match up as close as possible, then clamp them together. Step drill through both at the same time to near size and then ream with a 9/16" reamer.

Insert a short piece of 9/16" round rod in the reamed holes. Align the two pillars and clamp them in the milling vise and mill the bottom of the pillar base. Remove only enough material from the bottom of the pillars to make them even and flat. When finished, the distance between the bottom of the base and the center of the 9/16" hole should be 6-9/16" (+ or - 1/32").

Make the crankshaft bushings from 5/8" brass round rod. Turn the outside

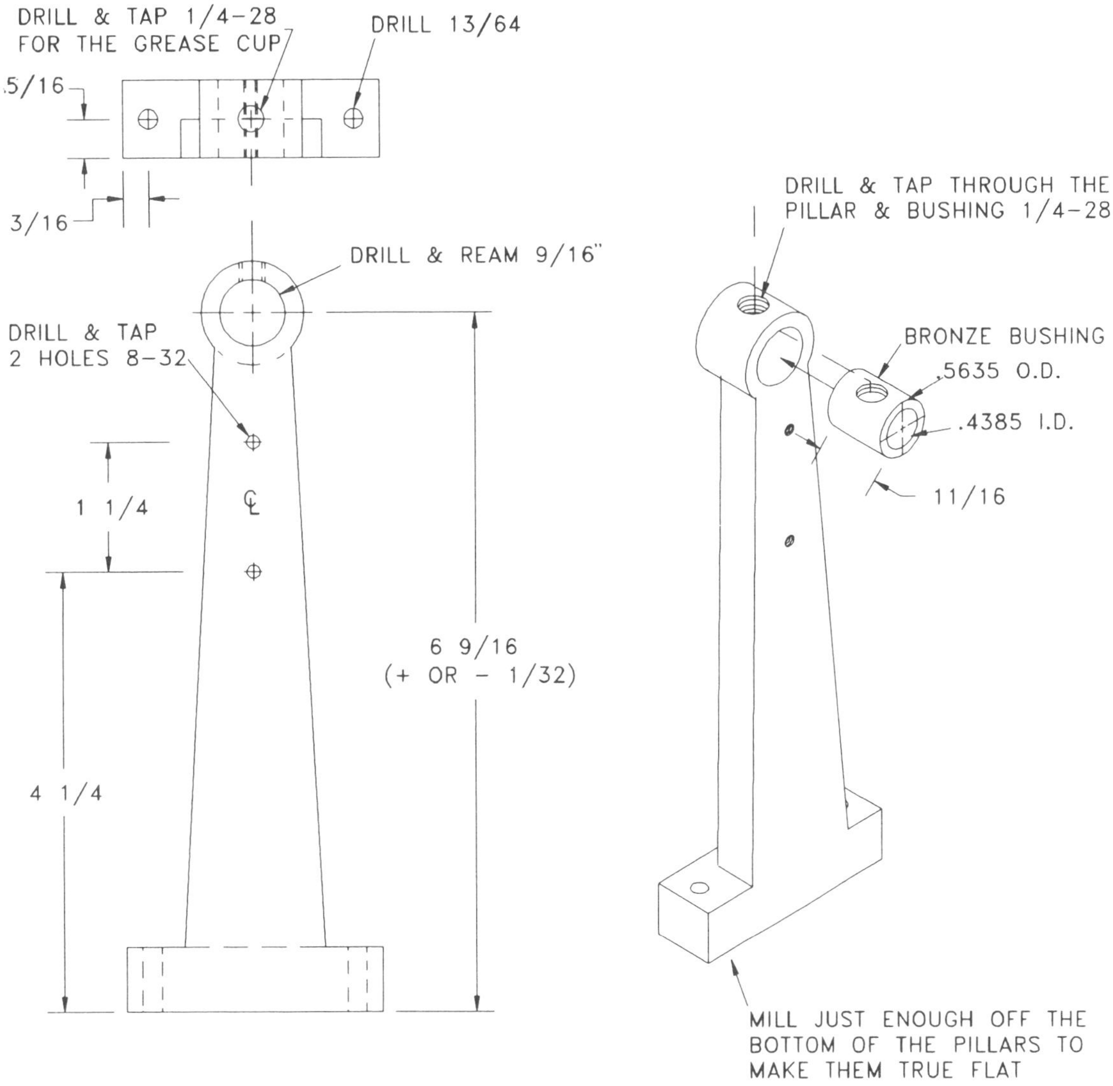

Figure 23 The crankshaft support pillar. Make an opposing pair

diameter to .5635". Drill the inside diameter near size and then ream .4385". Press a bushing into each pillar. The fit of the bushing should be a nice easy press fit.(Not more than .001" larger than the bore). If it's too tight the pillar could crack or break as the bushing is being pressed in. Drill and tap two holes 8-32 in the outside pillar where shown. These holes will be for the ignition plate.

It's important to drill the mounting holes in the bottom of each pillar straight. If you happen to have a drill press with a tilt table you can tilt the table over 90 degrees and position the pillar on the drill table so that its bottom is perpendicular

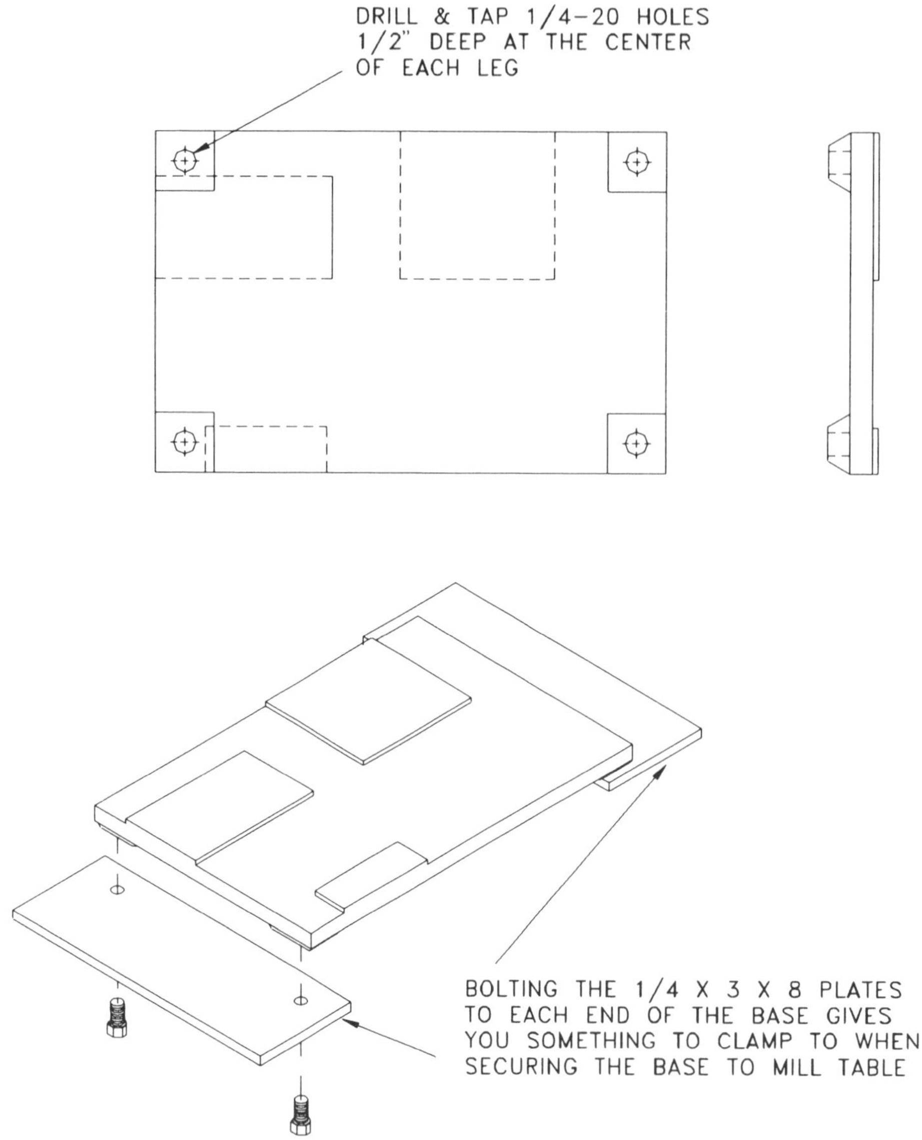

Figure 24 Setup for milling the base

Figure 25 **Milling the base**

with the drill spindle. Align the drill with the marked location of a 13/64" hole. Clamp the pillar in position with C-clamps and drill the hole. Reposition the pillar and drill the other 13/64" hole. To drill the 1/4-28 hole in the top of each pillar for the grease cups reposition the pillar so that the top is perpendicular with the spindle. Clamp it in position and drill and tap the hole.

If you don't have a tilt table on your drill press an angle plate mounted on the mill or drill table would work as well.

THE BASE: It is important to mill the top of the base flat so that the crankshaft support pillars and other components will align properly during assembly.

Mill the bottom of the base first. To set the base up for this operation, clamp it to the mill table face down so that the legs are facing up. You will have to use spacer furniture to make the base level on the table and to keep it from rocking back and forth. Use hold down clamps to secure the base to the mill table and make sure the base doesn't warp as you tighten the clamps. After the legs have been milled flat, drill and tap 1/4-20 holes in the center of each leg.

Make mounting brackets from 1/4" x 3" x 8" C.R.S.. Drill holes in the mounting brackets to align with the holes in the legs of the base. Bolt the brackets to the base with 1/4-20 x 1/2" bolts. Clamp the base to the mill table with hold down clamps so that the top side is facing up. Depending on how wide your mill table is, you may need another 1/4" spacer to raise the bolt heads off the table. Figure 24 shows the setup and figure 25 contains a photo of the operation. Use a flycutter or face mill to mill the surface of the mounting pads flat.

MOUNTING THE CRANKSHAFT PILLARS: It is impossible to give an exact mounting location on the base for the crankshaft pillars, pivot post and water jacket. There will be small differences in the sizes of the different parts and drilled hole locations will vary slightly between individual builders. I have however, tried to help a little by giving approximate mounting locations as laid out in figure 26. These are for your reference only and should not be drilled. Each of these parts will be located during construction.

To keep both crankshaft pillars in perfect alignment during mounting, insert a

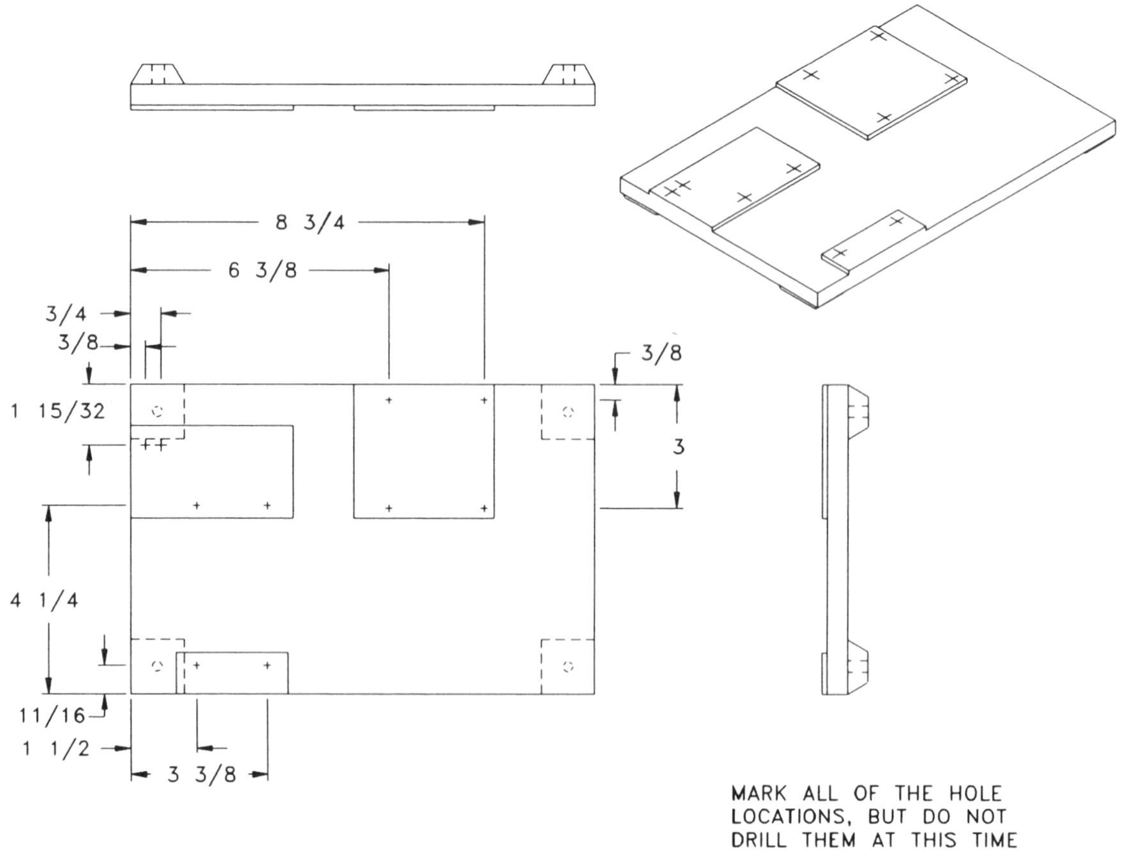

Figure 26 **Laying out the base**

7/16" shaft through the bushings in both pillars. During the mounting procedure check the shaft frequently to make sure it turns freely without binding.

Set the pillars with the shaft intact on the base mounting pads. Align them with the marked hole locations as close as possible. The outside pillar should be centered near the inside edge of the outside pad. To make sure that the pillars are aligned and that the shaft continues to turn freely, locate and drill the mounting holes in the base one at a time. For instance, start with the outside pillar and mark one of the mounting hole locations with a transfer punch. A transfer punch is a punch that allows you to locate a hole in relation to another hole. In this case we want to locate holes in the mounting base to align with holes in the crankshaft pillars. You can buy transfer punches or you can make your own out of drill rod. In this case we need a

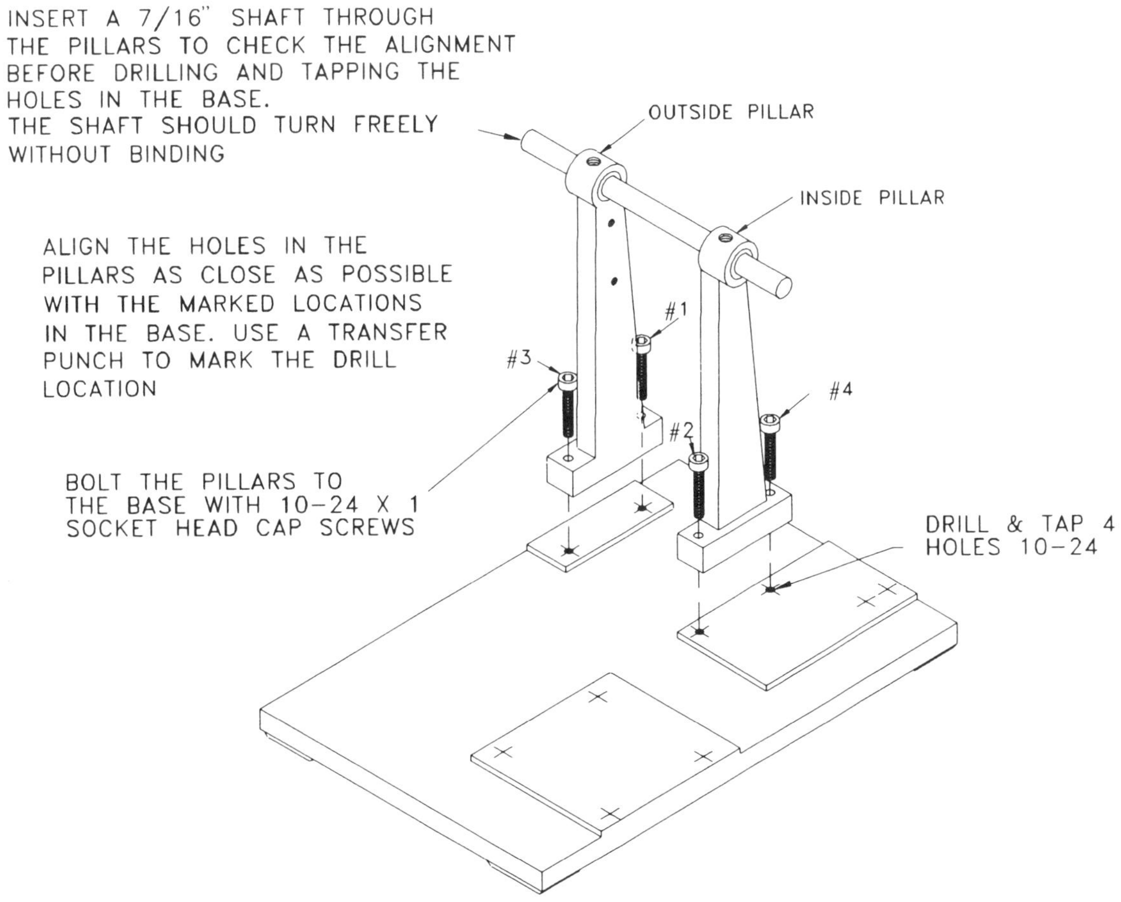

Figure 27 Bolting the crankshaft support pillars to the base

13/64" transfer punch. If you decide to make a punch, you probably will not have drill rod that size so you'll have to chuck a larger piece in the lathe and turn it to size. Turn a small point on the end and then harden the end by heating it to a bright red and quenching it. Temper the end by reheating it to a straw color and quenching. With the crankshaft pillar set in its desired position, insert the transfer punch through the base hole in the outside pillar designated #1 in figure 27. Mark its location by tapping the end of the transfer punch with a hammer. Drill and tap the hole 10-24 and bolt the pillar in position. Check the shaft for alignment and make sure it turns freely then locate hole #2 for the inside pillar. Drill and tap 10-24 and bolt the pillar in position. Check for alignment and make sure the shaft still turns freely and locate, drill and tap holes #3 and #4 one at a time. Finish bolting the pillars to the base. If the base or the bottom of the pillars are not true flat the crankshaft may bind. If this should happen, a solution would be to shim the pillars as required.

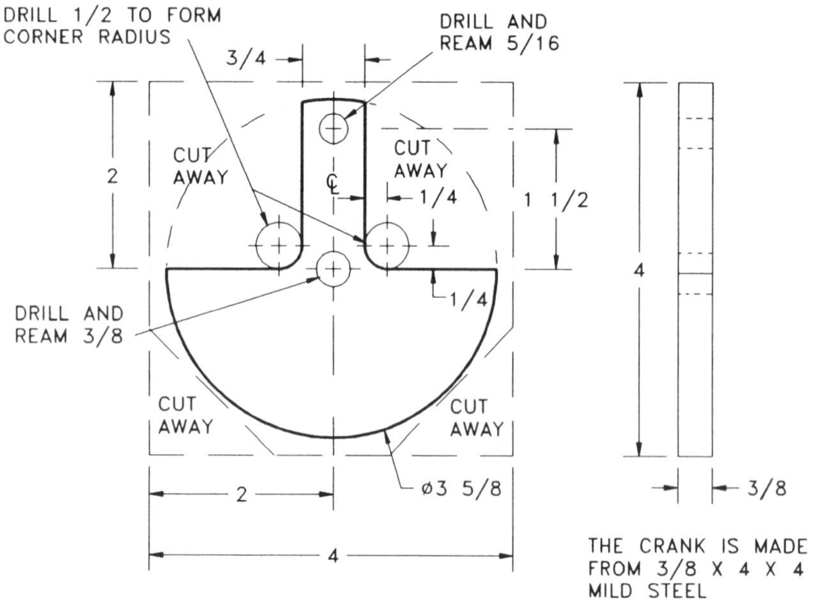

Figure 28 Layout for the crank

sequence is followed. Laying out the crank is very simple, but the holes must be located exactly because any difference here will cause the crank linkage to be out of wack. Make the crank from a piece of 3/8" x 4" x 4" mild steel. Figure 28 shows the layout of the blank. Find the location for the center of the blank and scribe the 1-13/16" radius, then lay out the crank end. Form the 1/4" radius at the inside corners by drilling with a 1/2" drill. In order to insure accurate holes and hole locations the following holes need to be step drilled. Step drilling means beginning with a small pilot drill and enlarging the hole with progressively larger bits. Locate the outside hole for the crankpin exactly 1-1/2" from the center. Step drill and ream it through 5/16". Step drill and ream the center hole 3/8". Cut the shape of the crank out with a hacksaw or bandsaw. The outside radius of the crank will be turned later on the lathe and the inside edges can be cleaned up with a file.

THE CRANKSHAFT AND CRANK: From now on subsequent parts will be located in relation to the crank so a definite

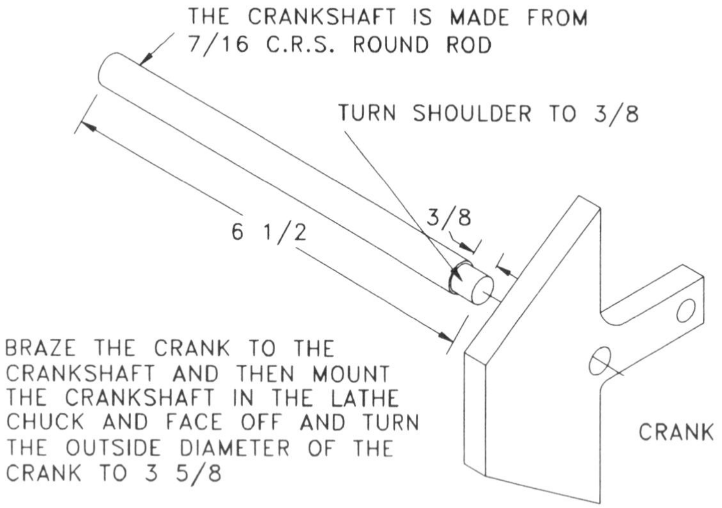

Figure 29 The crankshaft and crank

The crankshaft is shown in figure 29. Make it from a piece of 7/16" C.R.S. round

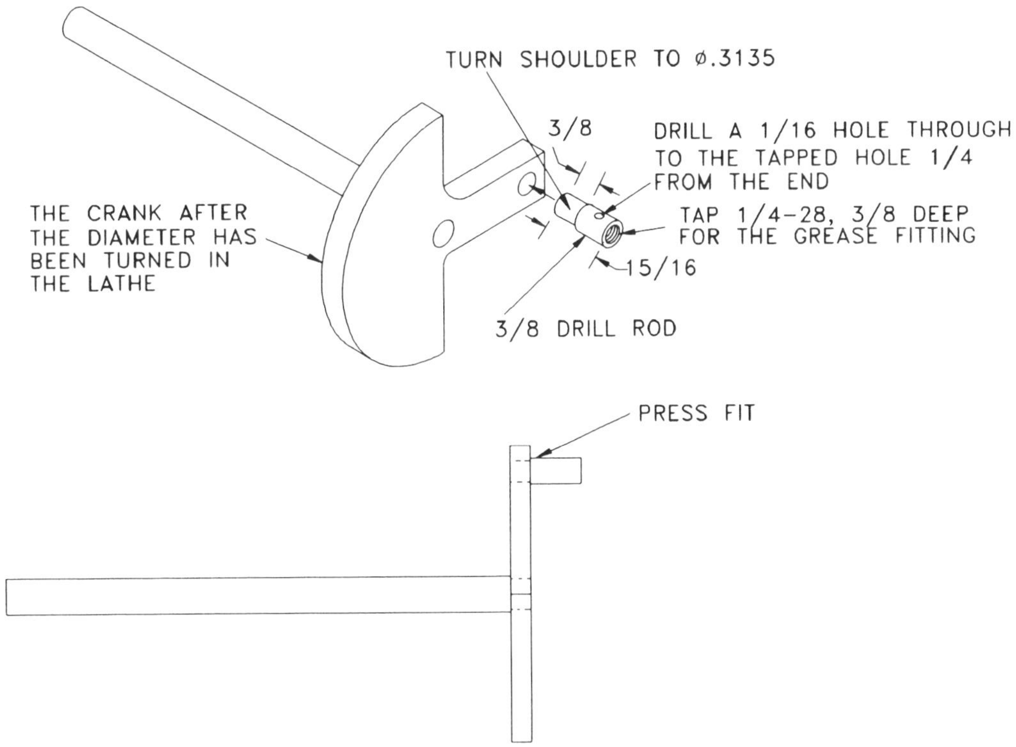

Figure 29A **Making and installing the crankpin**

rod 6-1/2" long. Turn the 3/8" diameter shoulder 3/8" back from one end.

Insert the 3/8" crankshaft shoulder end into the crank and braze it in place as shown in figure 29. Mount the crankshaft in a 7/16" collet and turn the outside radius of the crank and face it off. Clean up any sharp or rough edges with a file.

The crank pin and its installation is shown if figure 29A. It is made .001" oversize to press fit into the 5/16" hole in the crank. Make it from a piece of 3/8" round drill rod. Drill and tap one end 1/4-28, 3/8" deep. Chuck the other end of the pin in a collet and turn a .3135" diameter shoulder 3/8" back. Drill a 1/16" hole through to the 1/4-28 threaded hole 1/4" from the end. This hole lets the grease in to lubricate the connecting rod bushing.

MAKING THE CAMS: We will be making three cams. They are the exhaust, intake and ignition cams. Often in projects of this type you are given complicated instructions for making the cams. A rotary table on a milling machine is also often required. Well, we couldn't afford the rotary table and to get around buying one my Dad made what I think is a pretty neat fixture. Before we get into that we better make the cam blanks. The intake cam is shown in figure 30. The exhaust cam is shown in figure 31 and the ignition cam is shown in figure 32.

To make the blanks for the exhaust and intake cams mount a 1-7/8" length of 7/8" C.R.S. round rod in the lathe with 1-1/4" exposed. Reduce the diameter to .785". Center and step drill through to 7/16". Ream to .4385". With a 3/32" parting tool

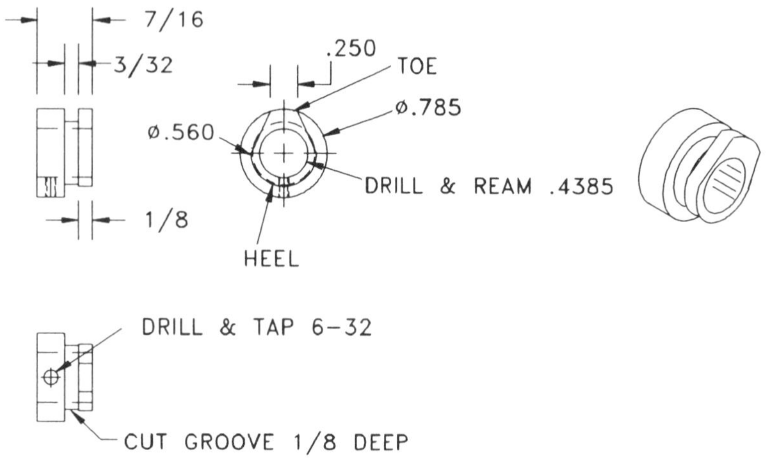

Figure 30 The intake cam

cut a groove .125" deep, 1/8" from the end. Part off at 7/16" in length. Repeat for the second cam.

To make the blank for the ignition cam, turn the end diameter of a .785" blank to .625", 1/2" back. Center, step drill and ream to .4385". Part off at 3/4" length. Drill and tap all 3 blanks as shown for 6-32 set screws.

The fixture used to make the cams is shown in figure 33. In operation the stationary arm of the fixture is held in the mill vise. The cam blank is placed on the end of the shaft and secured with a set screw. To form the heel of the cam, the pivot arm is moved in a direction to feed the work against the rotation of the milling cutter. As each cut is made the table is advanced to make the next cut. The sequence is shown in figure 34.

To form the toe of the cam the pivot arm is locked in position and a series of straight passes are made against the milling cutter. See figure 35.

Make the parts of the cam fixture as shown in figure 33. The pivot arm is held in place on the shaft with a 6-32 x 3/4" socket head cap screw. Allow about 3/4" of the shaft to extend out past the pivot arm. A 10-24 x 3/8" socket head cap screw screwed into the end of the shaft will keep it from slipping out of the stationary arm. The tension of the shaft rotation is controlled by the tightening the 10-24 x 1" socket head screw on the clamp end of the stationary arm.

Begin the first operation by mounting the exhaust cam on the fixture as shown in figure 34. Tighten the set screw against the

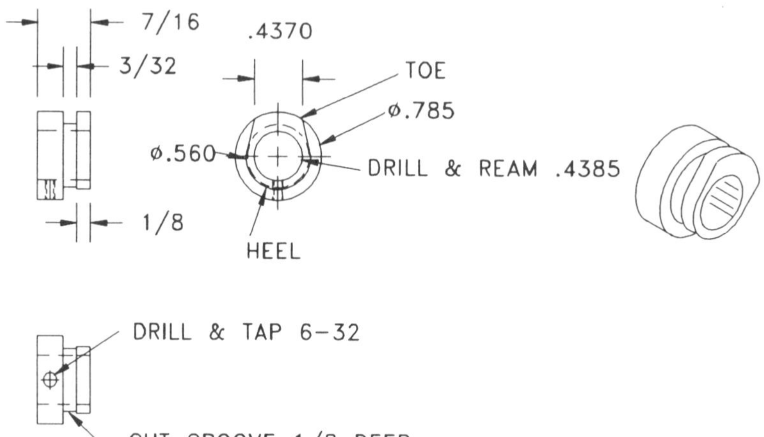

Figure 31 The exhaust cam

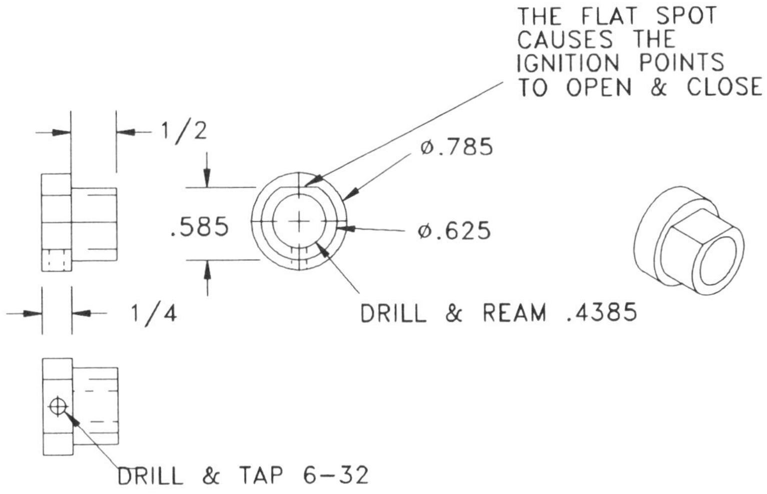

Figure 32 The ignition cam

setup shown is for a horizontal mill or a milling attachment on a lathe. On a vertical mill you would have to rotate the fixture in the vise so that the cam faced up.

The object is to reduce the heel of the cam to .560". Remember a milling cutter is dangerous and a careless maneuver can cause a serious injury. Make a couple of dry runs with the machine off to get familiar with the travel path making sure that there

flat on the shaft. Set the fixture up so that the front center of the cam aligns with the center of the side of the milling cutter. The are no obstructions in the path. Begin the first cut by pushing the handle of the fixture in a downward direction feeding the work

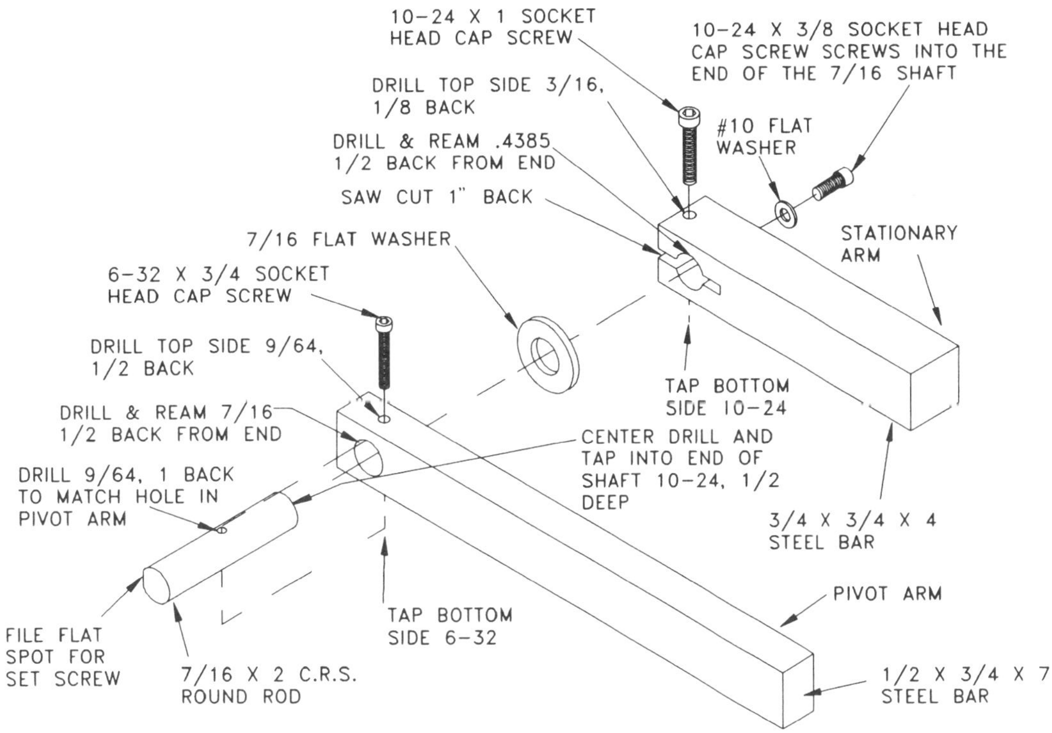

Figure 33 The cam fixture

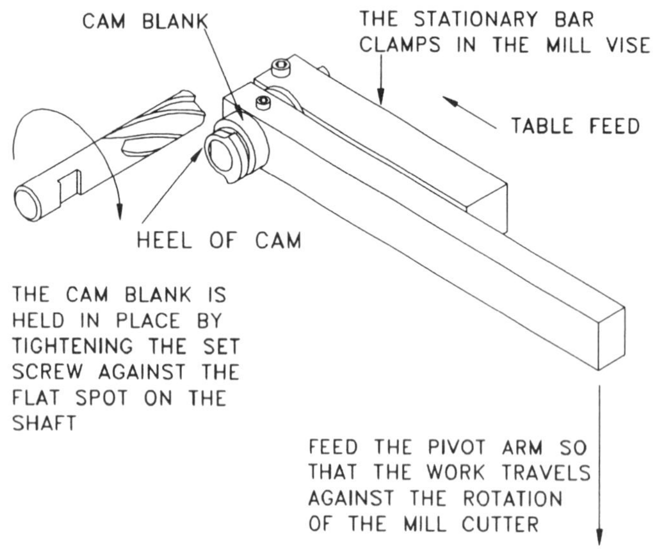

Figure 34 **Set up to turn the cam heel**

against the rotation of the milling cutter. Feeding in this manner makes it less likely for the cutter to grab the work out of your hands. Even so, there is still a danger of that happening so hold the handle securely and don't over feed or try to take off too much material at once. About .020" per pass is sufficient.

To form the toe on the cam it is necessary to change the position of the fixture in reference to the milling cutter. Align the bottom center of the cam with the top center of the milling cutter. Lock the pivot arm in position and feed the work straight in against the milling cutter. The object is to reduce the toe of the exhaust cam to .4370". Figure 35 shows the setup. A single pass of the cutter forms one side. Then reposition the work in reference to the cutter and make a series of passes on the opposite side of the cam to complete the form.

Make the intake cam next. The procedure is the same except that you want the toe of the intake cam to be .250".

The milled flat on the ignition cam is

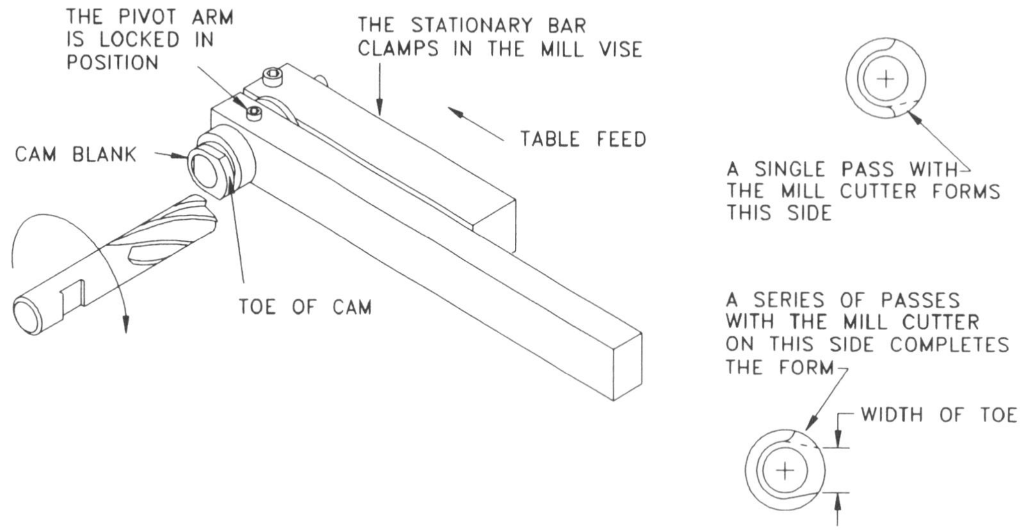

Figure 35 **Set up to turn the toe of the cam**

made on the same fixture. The purpose of the flat is to provide at least a .025" deflection of the points. Lock the fixture swivel by tightening the pivot handle clamp on the shaft. Feed the cam straight in to form the flat until the diameter at the flat is .585".

THE IGNITION MOUNTING PLATE: Make the plate from a piece of 1/16" x 2-1/4" x 3-1/4" sheet steel. It is laid out to mount a set of points from a late 1970's Ford V8. Many different types of points would work, but if you use a different type you will have to lay the plate out differently than is shown in the drawing. There are several holes drilled in the plate and their locations are shown in figure 36. The two holes tapped 8-32 are for mounting the points. There is a short stub located on the back of the points and it goes in the #25 hole. The points pivot at this location. The crankshaft fits through the 1/2" hole. The two holes that are drilled for an 8-32 C.S. screw are for mounting the plate to the outside crankshaft support pillar. The oblong hole makes it possible to adjust the gap of the points with a screw driver. The hole that is tapped 6-32 is for mounting the condenser bracket.

A tip for laying out the plate is to locate and drill the #25 hole for the point pivot first. Then check the location of the two 8-32 mounting holes. When their location is confirmed go ahead and drill and tap them for 8-32 screws. Drill and tap the rest of the holes as shown. The oblong hole

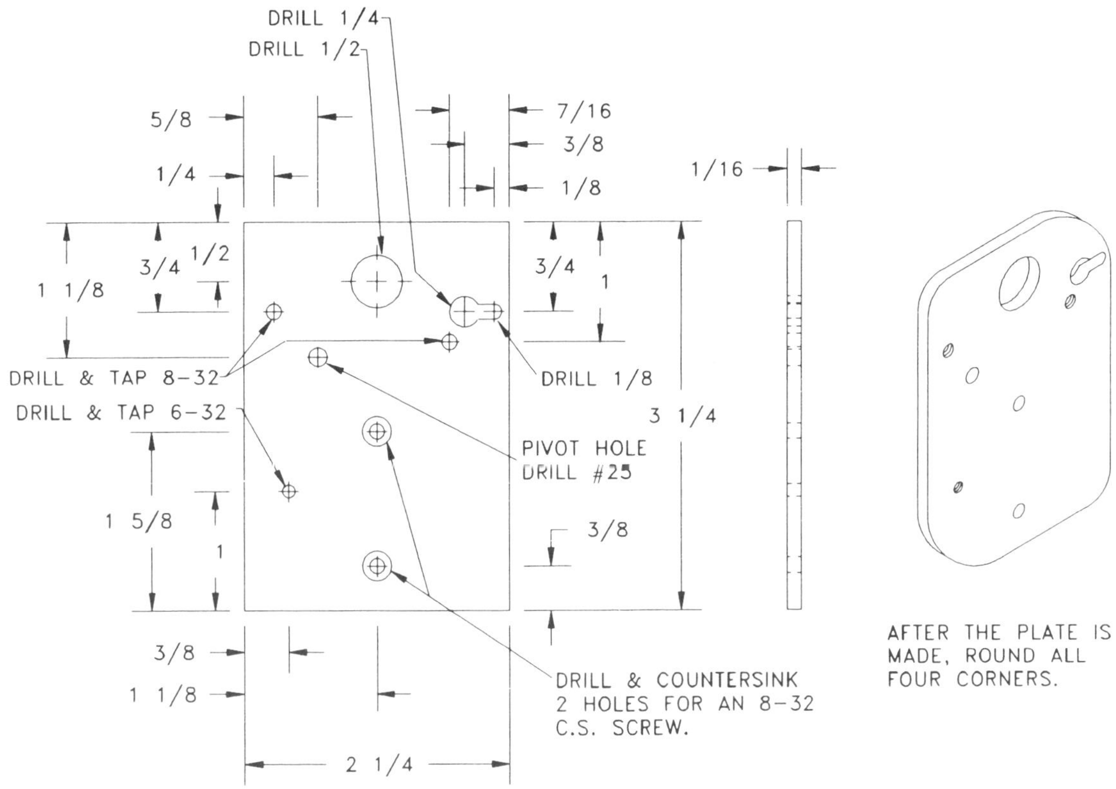

Figure 36 Ignition mounting plate

can be formed with a file. After the plate has been made, grind a radius on the corners to get rid of the sharp edges.

Mount the ignition plate to the outside crankshaft support pillar with two 8-32 x 1/2" C.S. screws.

INSTALL THE CRANKSHAFT: Slip a 7/16" ID x 3/4" OD x 1/8" long spacer bushing on the crankshaft and then insert the crankshaft into the inside support pillar bushing. As it goes through the bushing, slide the intake cam, a 7/16" friction washer, the exhaust cam and the ignition cam onto the shaft. Figure 37 shows how they should look as they go on. The brass friction washer mentioned can be found in the plumbing department at the hardware store. It is intended for use with the stem packing in valves and faucets.

MACHINING THE FLYWHEEL: Before machining the flywheel look at figure 38, 39 and 40. One of the things we decided when we built our engine was to have the back curve of the spokes face the direction

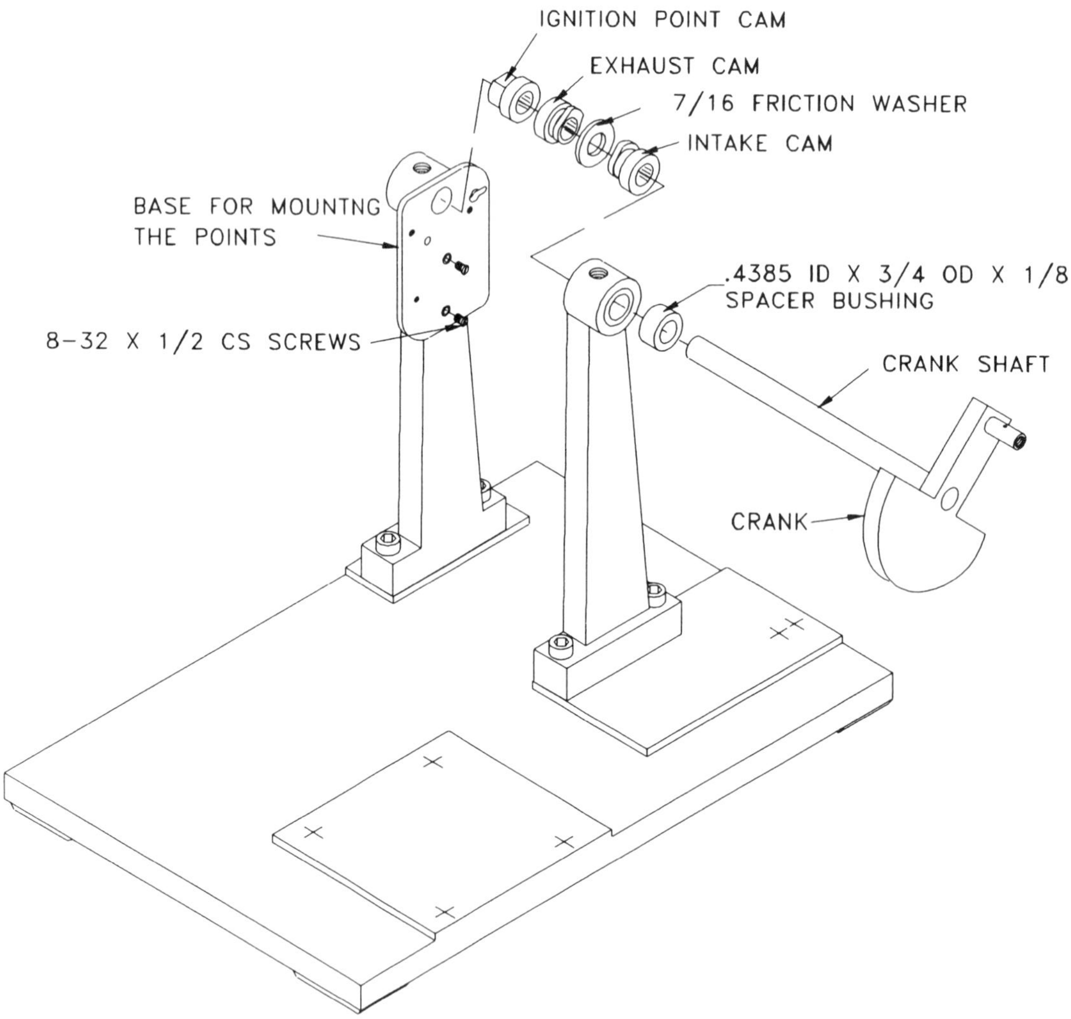

Figure 37 **Mounting the cams, crankshaft and ignition plate**

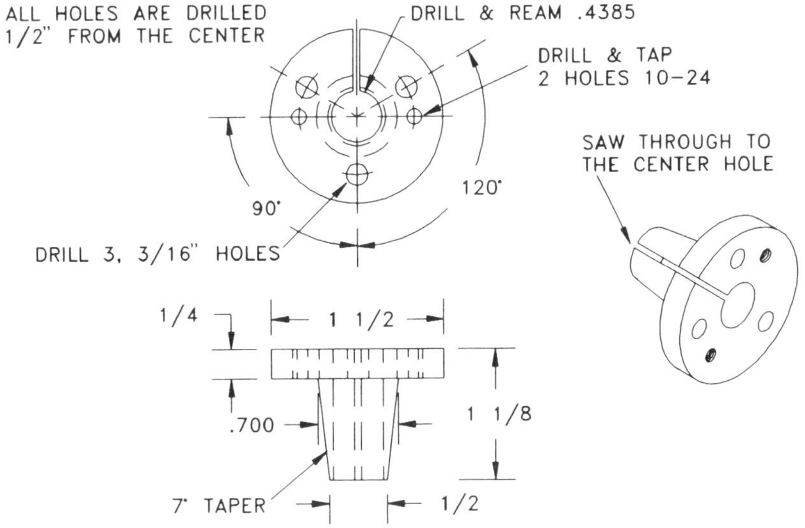

Figure 38 The flywheel hub insert

of rotation. I have seen it both ways and I am not sure if it really matters or not. Anyway, that's the basis we used in determining the front and back of the flywheel. The sides are labeled in figure 39.

The front side of the hub has a tapered bore. A split tapered hub insert fits into the tapered bore. As the insert is tightened it clamps down on the crankshaft and secures the flywheel to the shaft. This method of holding the flywheel on the shaft works very well and is certainly much better than set screws which always seem to work loose. Another nice feature is the ease in which the flywheel can be removed. It's simply a matter of removing the bolts and screwing one into each of the threaded holes in the insert. As the bolts are tightened they push against the flywheel hub and force the insert out loosening the flywheel.

To make the tapered hub insert, chuck a 3" length of 1-1/2" C.R.S. round rod in a 3-jaw chuck. Center drill, drill to near size and ream .4385". Turn the .700" shoulder 7/8" back. Adjust the compound rest and turn the 7 degree taper. Part off the insert 1-1/8" long. Layout and drill the three 3/16" mounting holes 1/2" from center and 120 degrees apart. One way to divide the circle in three equal parts is to wrap a piece of masking tape around the outside diameter. Trim the tape to fit exactly. Remove the tape and divide it into 3 equal sections marking each section. Rewrap the tape around the outside diameter. Use a tri-square and the marks on the tape as a reference and mark the location of each of the three holes.

Saw through one side of the hub insert where shown in figure 38. Locate, drill and tap the two threaded holes in the hub insert 10-24. They are located on the center line and perpendicular to the saw slit.

The flywheel drawing is shown in figure 39. Begin the machine work by chucking the front side of the flywheel in a 4-jaw chuck. Center it as close as possible. Face off and turn the outside diameter of the hub. Drill through 3/8" and then counter bore to 1" diameter 5/16" deep.

Remove the flywheel and rechuck it from the rear, center it in the chuck and face off and turn the outside diameter of the hub. Increase the inside diameter to 1/2" and bore the 7 degree taper.

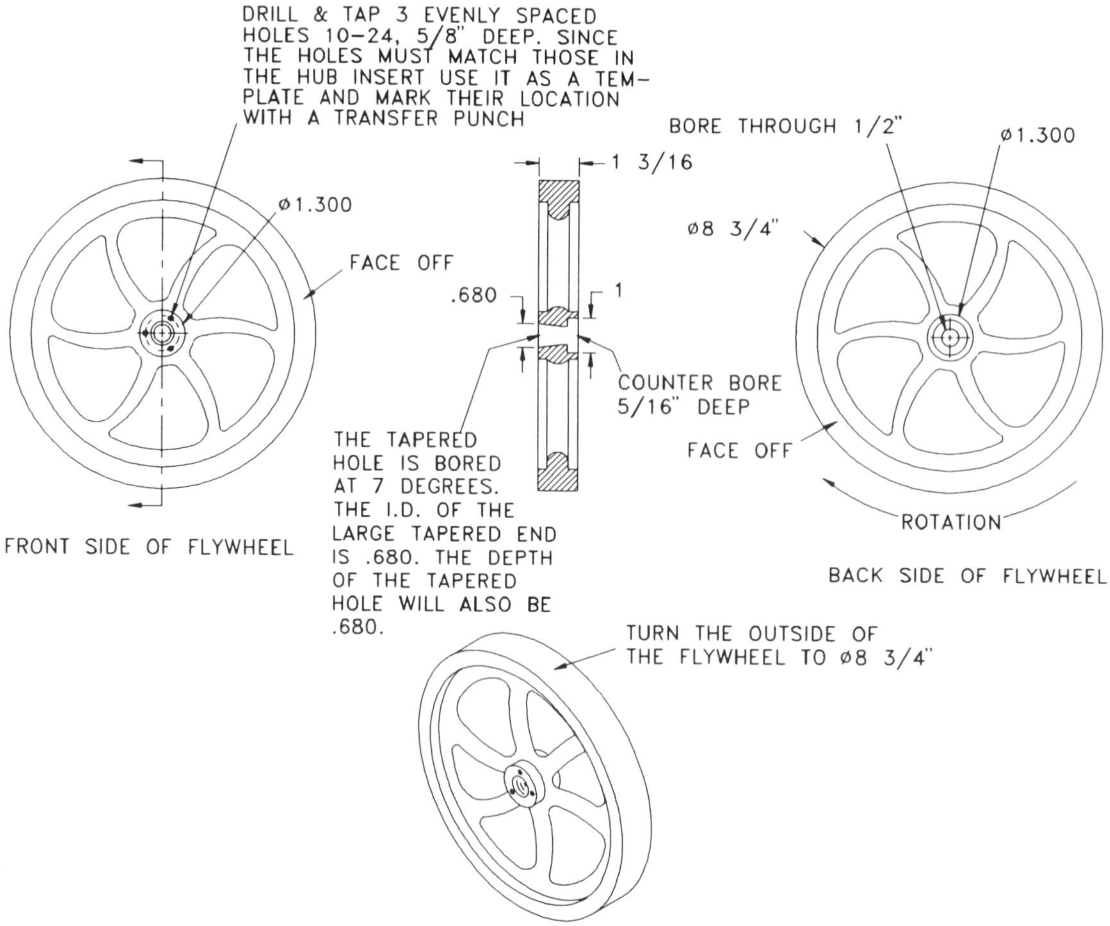

Figure 39 The flywheel

To locate the bolt pattern on the hub of the flywheel, use the hub insert as a template. Mark the hole locations with a transfer punch. Drill and tap them 10-24, 5/8" deep.

To machine the outside diameter of the flywheel it is easiest to prepare an arbor for between the centers work. To make the arbor, chuck a 5/8" shaft with about 5" protruding. Center drill it with the tail stock chuck and support it with a tail center while reducing to 7/16". Mount the flywheel on the arbor using the tapered hub insert to secure it. Support the arbor with the tail center while machining the O.D. and sides of the flywheel.

Mount the flywheel on the crankshaft with 3, 10-24 x 3/4" socket head cap screws. See figure 40.

MAKING THE CONNECTING ROD AND LINKAGE: In this section we will be making the connecting rod, The pivot post, the pivot arms and the piston rod. Accuracy in the production of these parts is very important. We found that small differences

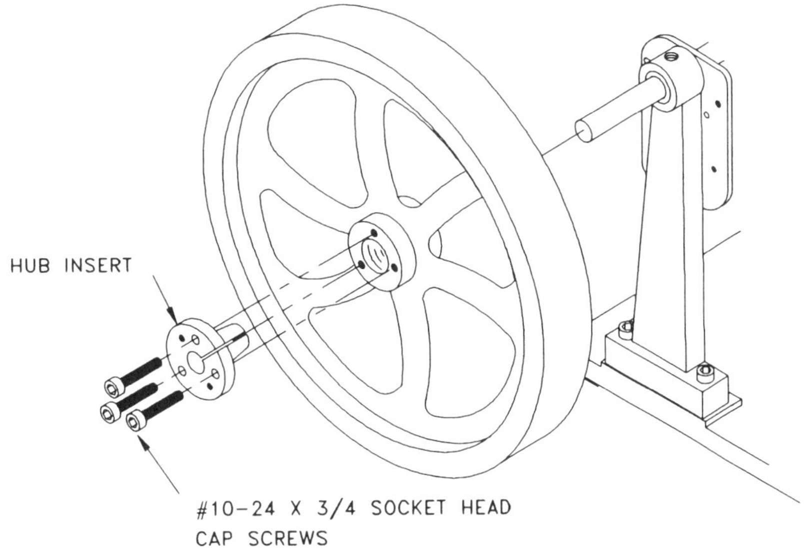

Figure 40 **Mounting the flywheel**

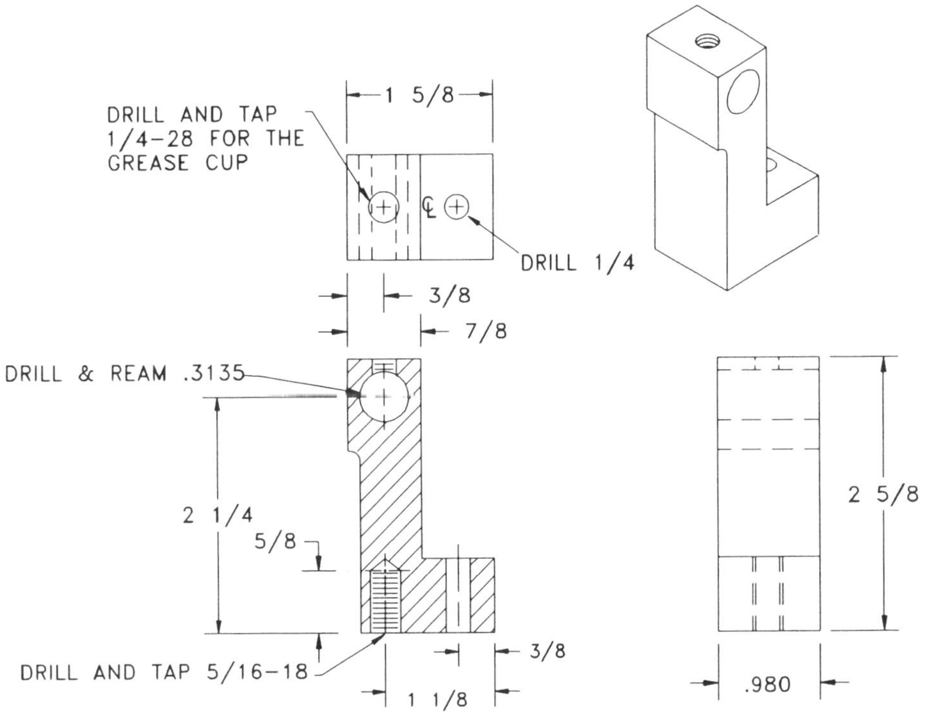

Figure 41 **The pivot post**

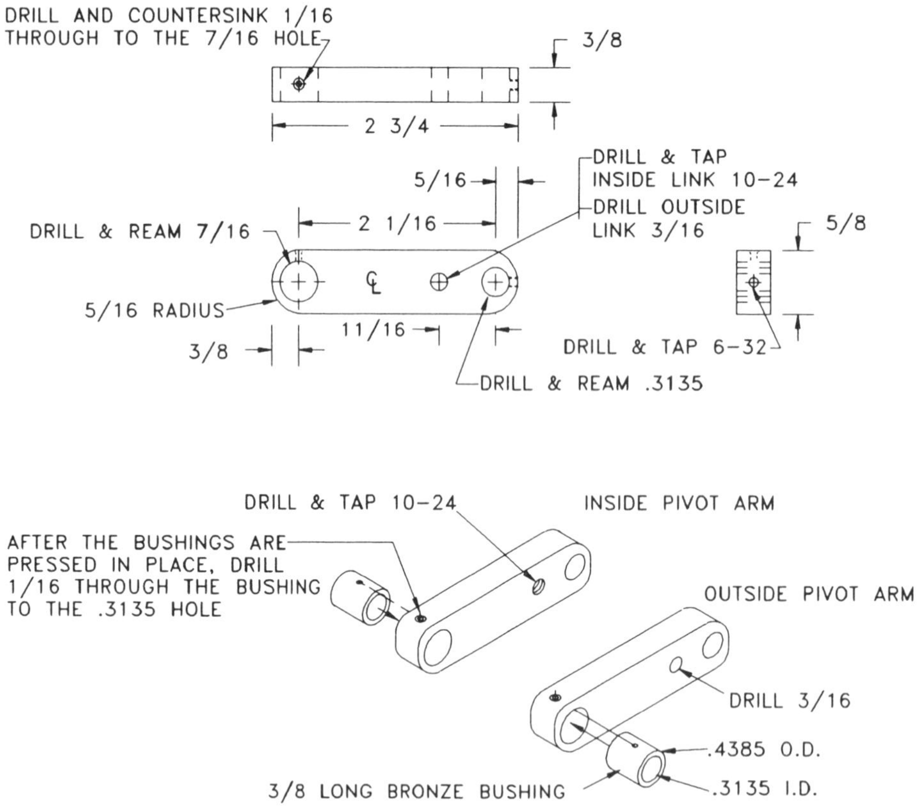

Figure 42 The pivot arms. Make two

in the locations of holes in these parts made a drastic difference in the linkage travel path.

Make the pivot post first. The layout is shown in figure 41. Mill the bottom of the post flat. Mill both sides to a finished width of .980". Drill and ream the .3135" hole for the pivot bushing. Drill and tap the 5/16-18 hole and drill the 1/4" mounting holes in the bottom of the post. Drill and tap the 1/4-28 hole for the grease cup in the top of the pivot post through to the .3135" hole.

The pivot arms are shown in figure 42. They are made from 3/8" x 3/4" aluminum bar stock and are 2-3/4" long. Begin by milling to size 3/8" x 5/8". These parts must be an exact match so it is best to clamp them together when drilling and milling them. In operation, the two links will be held together with a #10-24 socket head cap screw. Locate and drill for the cap screw holes first. The cap screw slips through the outside link so drill a 3/16" hole in it as shown. It will thread into the inside link so drill and tap it 10-24 in the location shown. You can now hold the two links together with a 10-24 screw while you drill the other holes and mill the radius on the ends. Drill the 7/16" hole for the bushing. Drill and ream the .3135" holes. Drill and tap a 6-32 hole in the back of each link through to the .3135" hole. These holes are for the set screws that will hold the pivot pin in place.

Make the two .4385" O.D., .3135"

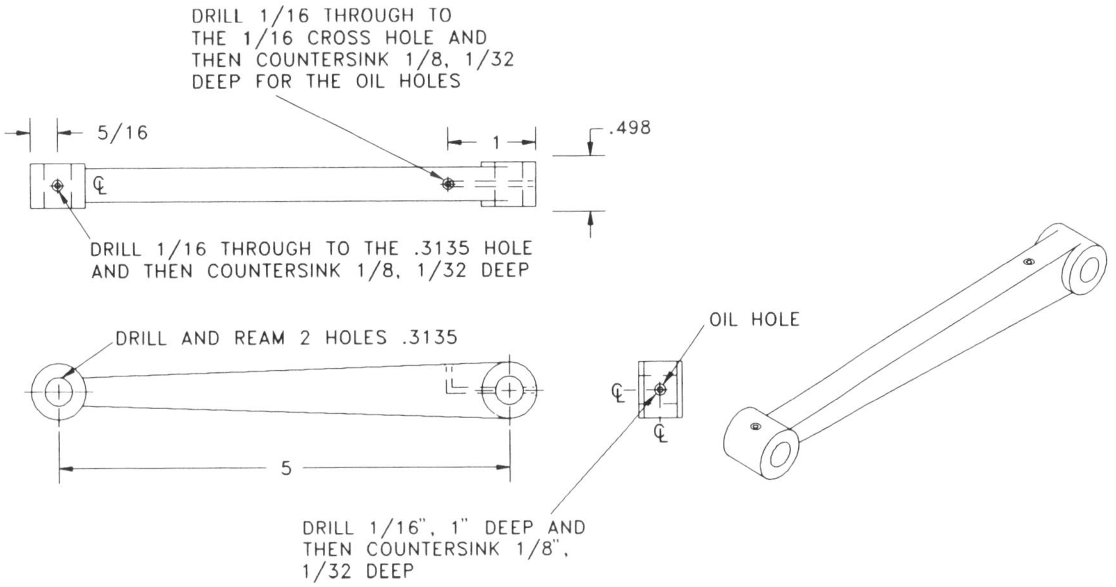

Figure 43 Piston rod

I.D. x 3/8" long bronze bushings. Press them into the 7/16" holes in the pivot arms. Countersink and drill a 1/16" oil hole into the top of each link and through the bushing to the .3135" hole. The radius on the ends of the links can be formed with a file or you can mount them on the cam fixture we made earlier and turn the radius on the milling machine. The smaller holes in the pivot arms will not fit the shaft on the cam fixture, but it is a simple matter to make another shaft with an end sized to fit the .3135" holes.

The piston rod is shown in figure 43. Mill the sides of each end of the rod equally to a finished width of .498". It is not necessary to mill the center section of the rod, just the ends. Locate, drill and ream the two .3135" holes to near size and then ream to size. Finally drill the 1/16" oil holes as shown. One way to drill the 1/16" oil hole in the end is to center the piston rod in a 4-jaw chuck and drill with the tail stock.

The connecting rod is shown in figure 44. Mill the back, front and sides of the connecting rod. The finished outside width of the yoke portion of the rod should be .980" and the finished width of the rod portion should be .480". Locate, drill and ream the 1/4" and the .3125" holes in the yoke portion of the rod. Also drill and ream the 1/2" hole at the top of the rod. Clamp the connecting rod in the mill vise and adjust it to the proper angle to cut the 1/2" wide 60 degree slot that forms the yoke portion of the connecting rod. The depth of cut on the low end of the slot should be 7/8".

Make the 3/8" I.D. x .501" O.D. x 9/16" long bronze bushing. Press it into the connecting rod. Make the piston rod pivot pins from 5/16" round drill rod. Turn a

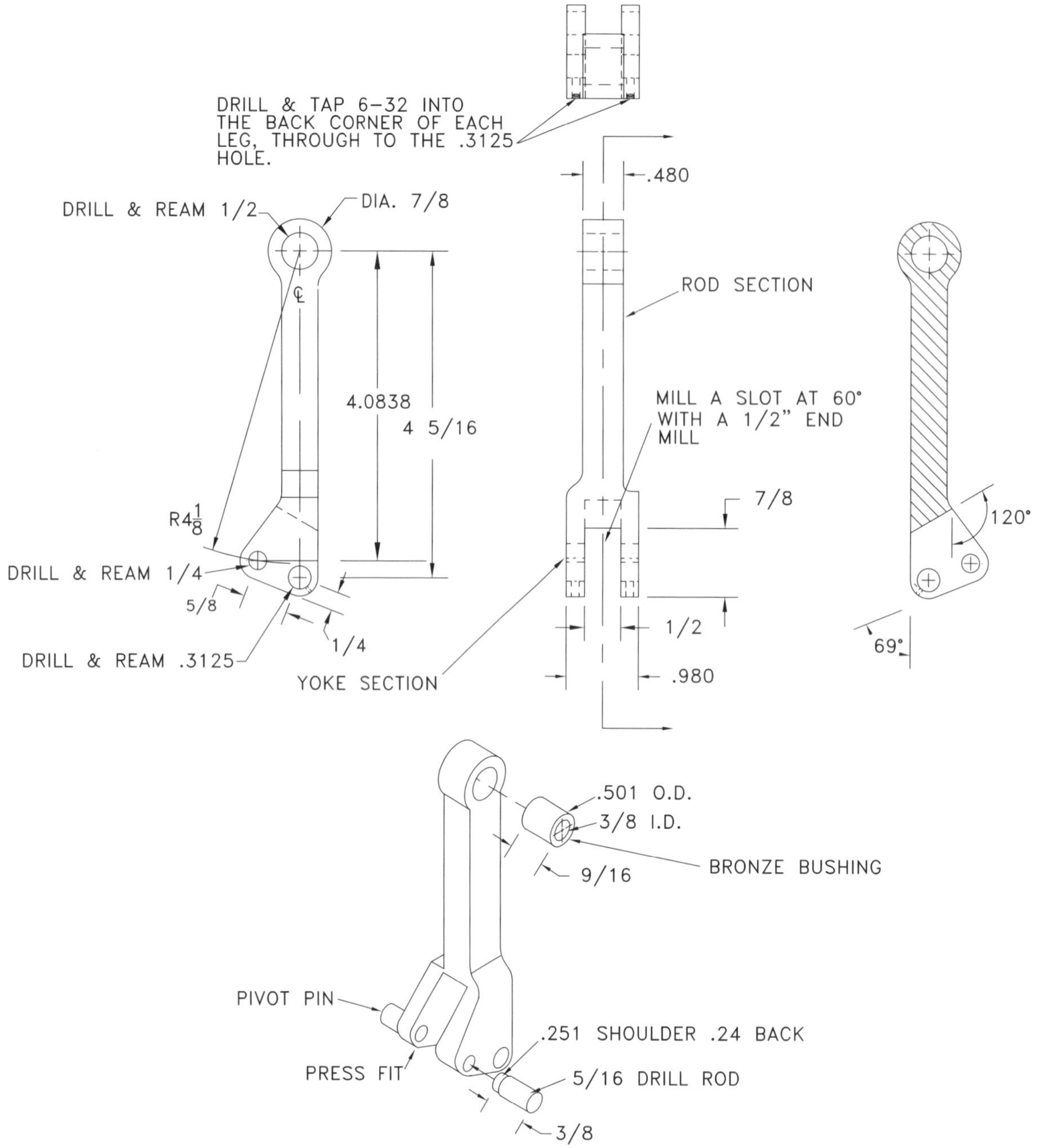

Figure 44 **The connecting rod**

.251" shoulder .24" back from the end of each one. Press the pins into the connectiong rod as shown. Set a piece of 1/2" material between the legs of the yoke to prevent them from collapsing while pressing in the pins. The pins and bushing are shown in figure 44.

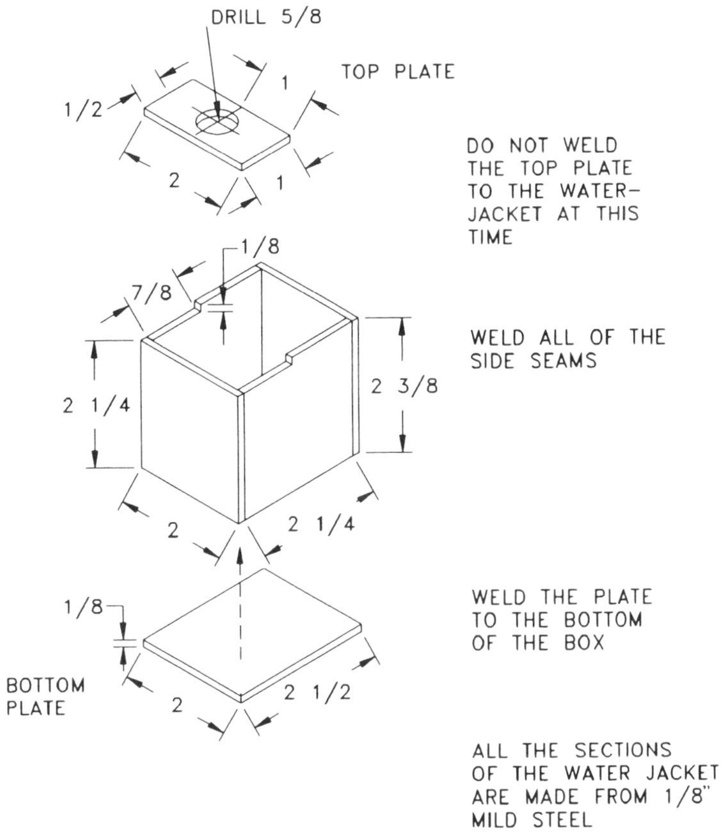

Figure 45 Water jacket

THE WATER JACKET: The water jacket is simply a box that holds the cylinder and the water to cool it. It is made from 1/8" sheet steel and is welded together. Figure 45 shows a drawing of the box and its individual parts. Half of the top of the box is left open. This allows the water to have an open surface area for evaporation which helps in cooling the engine. Even so the cooling water will reach a boiling temperature after about twenty minutes of running time. The 5/8" hole in the top cover of the water jacket is for the oiler adapter.

Cut all of the parts for the water jacket from 1/8" sheet steel. A piece of 1-1/2" angle works well as a jig to hold the sections of the water jacket together for welding. Clamp the angle in a vise. Clamp the bottom and a side section of the water jacket to the angle iron with C-clamps in such a way that the ends are held together. Weld the ends together and then rotate the box around and clamp the next side to the angle, align the edge, clamp in position and weld the seam. Place the bottom on the water jacket and weld it in place.

Bore the 1-1/2" hole through the water jacket for the cylinder where shown in figure 46. One way to do this, is to center the box in a 4-jaw chuck. Use furniture to create a space between the face of the chuck and the water jacket to pre-

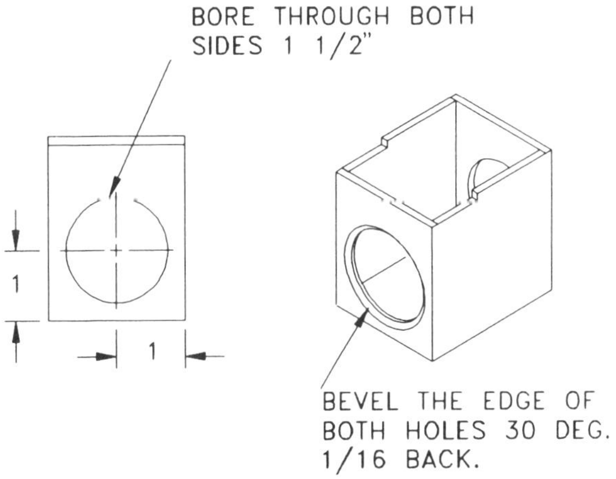

Figure 46 Boring the 1-1/2" hole in the water jacket for the cylinder

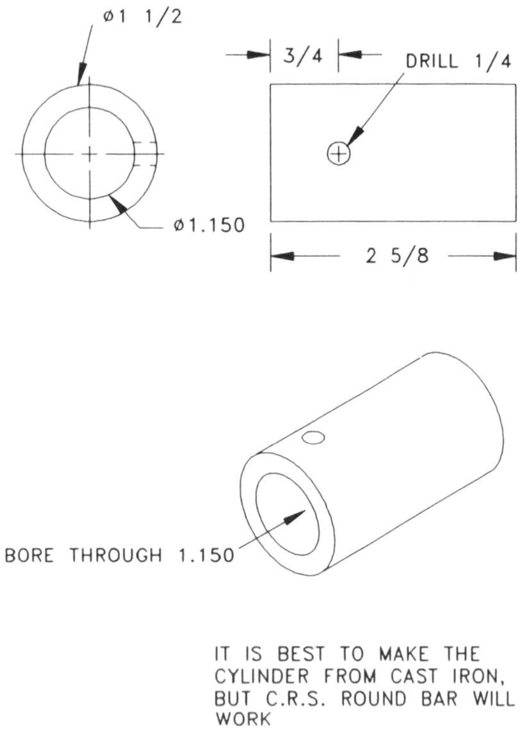

Figure 47 The cylinder

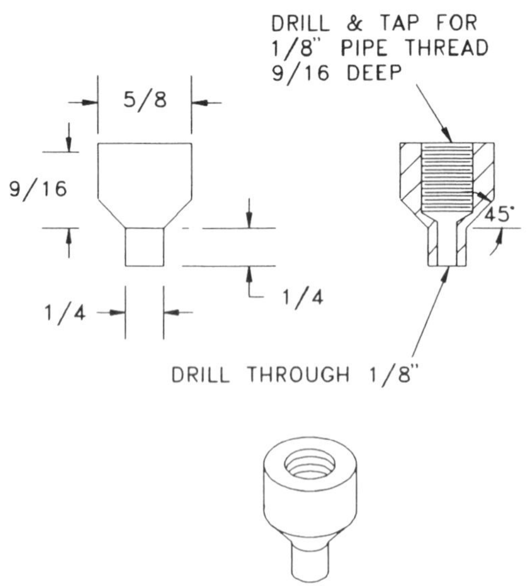

Figure 48 Oiler fitting

vent the boring bar from hitting and damaging the chuck as it comes through the back end of the jacket. Bevel the outside edge of each hole 30 degrees 1/16" back. This gives a fill area for the silver solder when the cylinder is soldered in place. Figure 46 shows the water jacket with the holes bored.

Make the cylinder next. It is best to make it from cast iron, but C.R.S. will work. Make it from 1-1/2" round stock. Chuck the cylinder in a 3-jaw chuck and bore to 1.15". The cylinder will be bored to its final size later after it has been soldered in the water jacket. Remove the cylinder from the lathe and mark the location for the 1/4" oil hole as shown in figure 47. It is best to clamp the cylinder in a V-block to hold it steady as you drill the hole.

The fitting for the oiler is shown in figure 48. Put a piece of 5/8" C.R.S. round rod 3-1/2" long in a 3-jaw chuck so that 1-1/4" of material sticks out past the chuck.

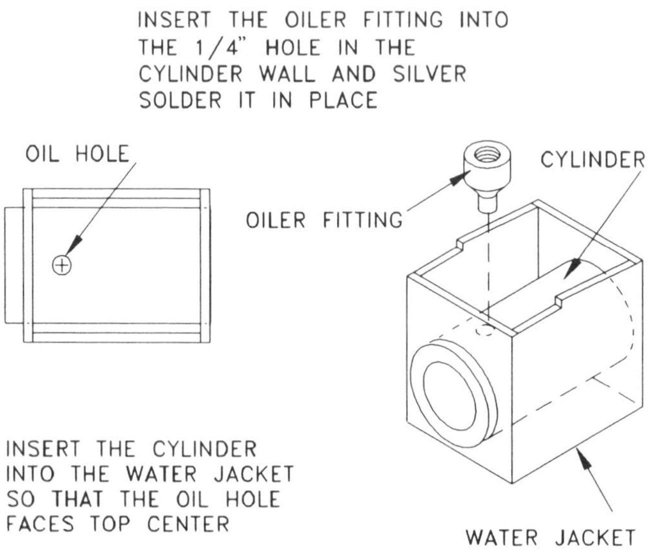

Figure 49

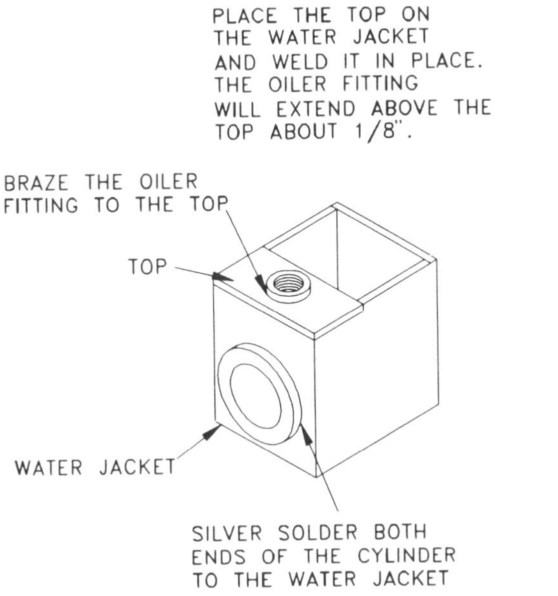

Figure 50

Face off, center drill and drill 1/8" through. Drill 11/32", 9/16" deep and tap for 1/8-27 pipe thread. Cut off to 13/16" in length. Chuck the threaded end and turn the 1/4" shoulder 1/4" back. Adjust the compound rest to turn the 45 degree taper.

Insert the cylinder into the water jacket as shown in figure 49. Notice that the oil hole in the cylinder is on the notched end of the water jacket. Also notice that the cylinder will extend out the back end of the water jacket about 1/8". Insert the oiler fitting into the 1/4" hole, apply flux and silver solder it in place.

Set the top on the water jacket as shown in figure 50 and weld it in place. Braze the oiler fitting to the top. The top of the fitting will stick up about 1/8" above the top surface of the water jacket. Apply flux around the cylinder joints at each end of the water jacket and silver solder the cylinder to the water jacket. After the water jacket has cooled, fill it with water and check for leaks.

Grind all of the weld seams on the water jacket smooth. A bench grinder with a coarse wheel will work fine for this purpose.

BORING AND LAPPING THE CYLINDER: Mount and center the water jacket in a 4-jaw chuck. Face off the head end and bore to 1.250". Remove and rechuck to face off the opposite end if necessary.

Lathes do not bore perfectly round holes and the tool finish left by a boring bar is not acceptable for a power cylinder bore. For that reason we must lap or hone the cylinder. Most shops do not have access to a pin hone so it will probably be necessary to make a lap. A lap is an abrasive charged tool that is applied to the work. It is made of material that is softer than the workpiece so that the abrasive grains will not be imbedded in the workpiece. Laps are commonly made of castiron, copper, lead, aluminum and hardwood.

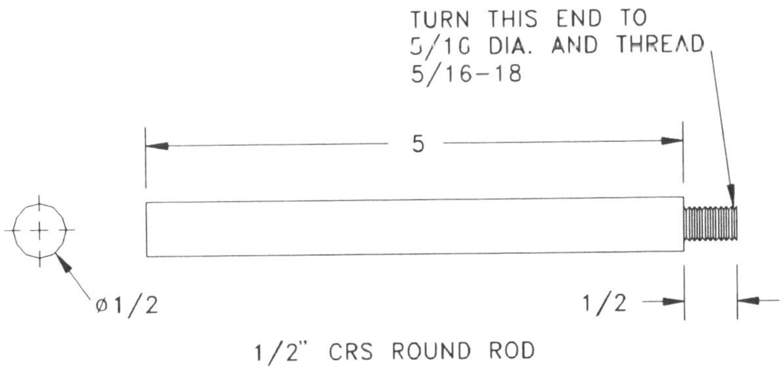

Figure 51 The cone mandrel

In this case we are going to make an expandable lap. Figure 54 shows an exploded view of the lap and figures 51-53 show the individual parts of the lap.

The cone mandrel is shown in figure 51 and is made from a piece of 1/2" C.R.S. round rod threaded 5/16-18, 1/2" back on one end.

The cone is shown in figure 52. To make it, chuck a piece of 1" diameter C.R.S. round rod in a 3-jaw chuck so that 1-1/4" extends out past the chuck. Face it off, center drill and drill and tap 5/16-18, 1" deep. Adjust the compound rest and turn the 7 degree taper 1" back. Part off at the 1" length. Cut the 3/16" wide x 3/32" deep slot with a hacksaw and clean it up with a file. You could also cut the slot with a metal slitting saw.

The expandable lap is shown in

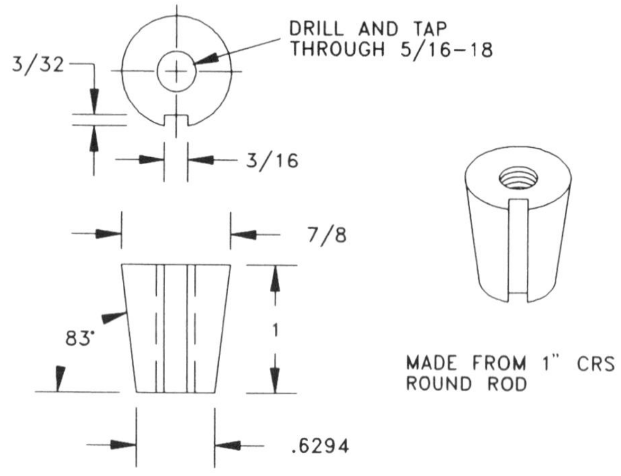

Figure 52 The cone

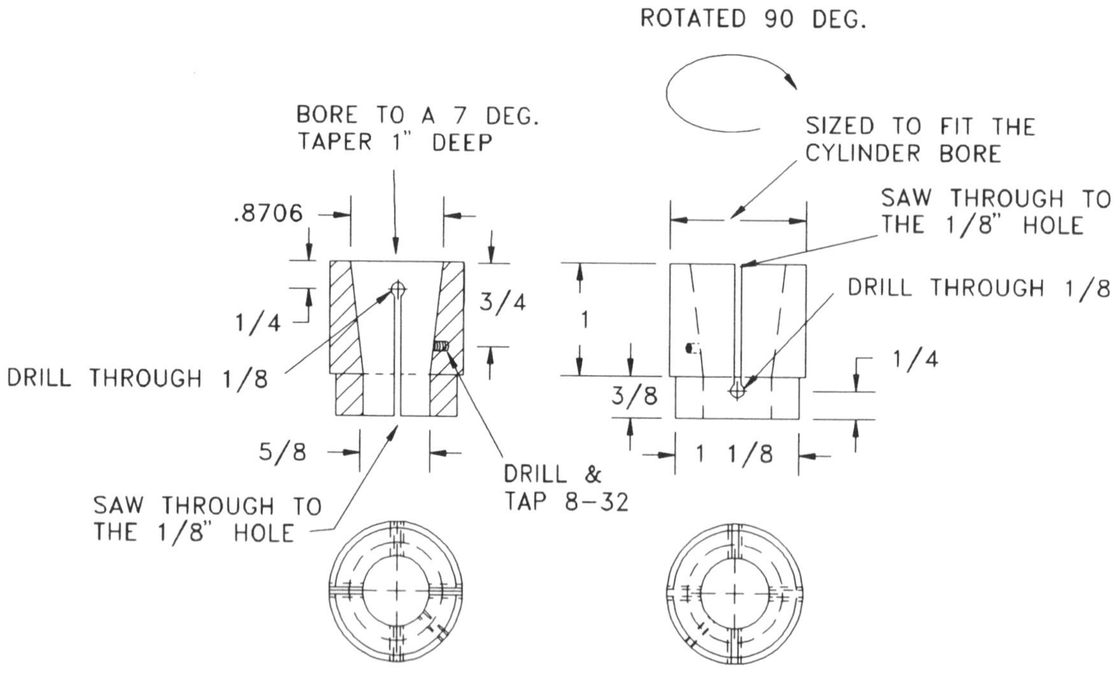

Figure 53 The expandable lap

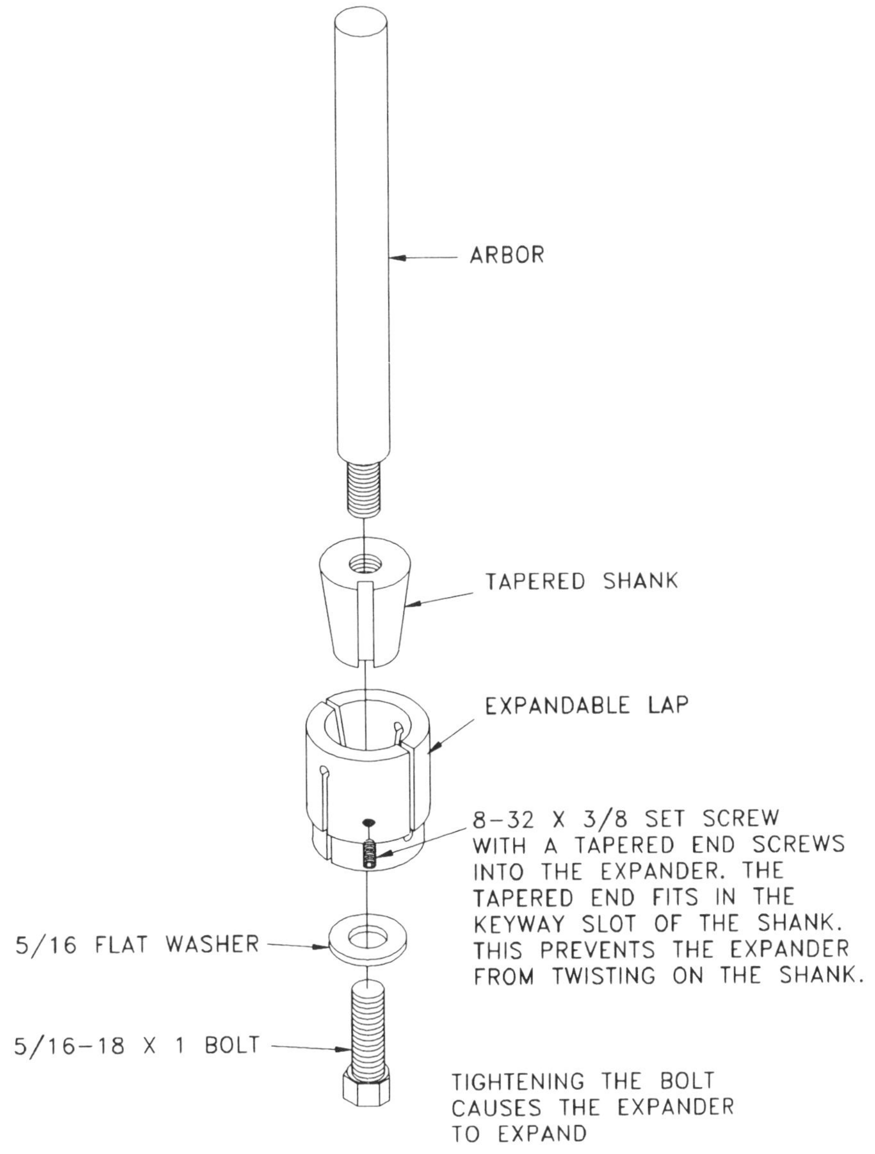

Figure 54 **The cylinder lap complete**

figure 53. It is made oversize and then reduced to fit the cylinder bore after it is mounted on the cone mandrel. To make it, chuck a piece of 1-3/8" x 4" aluminum round rod in a 3-jaw chuck. Turn the outside diameter to a size slightly larger than the cylinder bore. Face off, center drill and bore through 5/8". Adjust the compound rest and bore the 7 degree taper 1" deep. The large end of the taper hole should measure about .871". Cut the work off a little longer than 1" allowing for face off. Chuck the tapered end and face off and turn the 1-1/8" shoulder 3/8" back. Mark the locations for the 1/8" drill holes and the saw cut lines. Clamp the work in a V-block and drill the 1/8" holes. You can make the saw cuts with a hacksaw or bandsaw. Drill and tap the 8-32 hole for the set screw.

Assemble the cylinder lap as shown in figure 54. Tightening the 5/16-18 x 1" bolt causes the lap to expand. The 8-32 x 3/8" set screw with the tapered end screws through the expandable lap and into the slot on the tapered shank preventing the lap from twisting on the tapered shank during the lapping operation. After the lap is assembled, mount the mandrel in the lathe and reduce the diameter of the lap to fit the cylinder bore.

Before using the lap it must be charged with an abrasive compound. This is done by spreading a small amount of compound on a flat hard surface and rolling the lap through it. The abrasive particles embed in the softer aluminum. A grade 220 or 280 valve grinding compound will work fine.

Chuck the mandrel in the drill press. When lapping, you want to use a slower speed because at higher speeds it is hard to hang on to the cylinder and the lap has a tendency to grab it.

As the lapping tool revolves, flush the work by dripping kerosene through the cylinder with a small paint brush. Keep the cylinder moving in an up and down motion constantly to prevent RINGING, which is a deep groove in one place. To avoid BELL-MOUTHING, don't allow the lap tool to go more than 1/8" beyond either end of the cylinder. Bell-mouthing means a widening of the ends of the bore.

When you begin lapping you will feel more resistance in some areas than others. You continue lapping until these high areas are cut away. You can expand the lap from time to time by tightening the bolt at the bottom of the lap tool. You can tell when the lap needs recharging because it will loose its abrasive feel and no longer cut. Continue lapping until a visual inspection shows a uniform texture from end to end. The object is to produce a cylinder that is truly round with parallel walls. Your final finish will depend on the condition of the bore when you began. Lapping is a slow process and it can take hours to increase the bore of a cylinder a few thousandths of an inch. Tiny flaws or a slight ripple will not

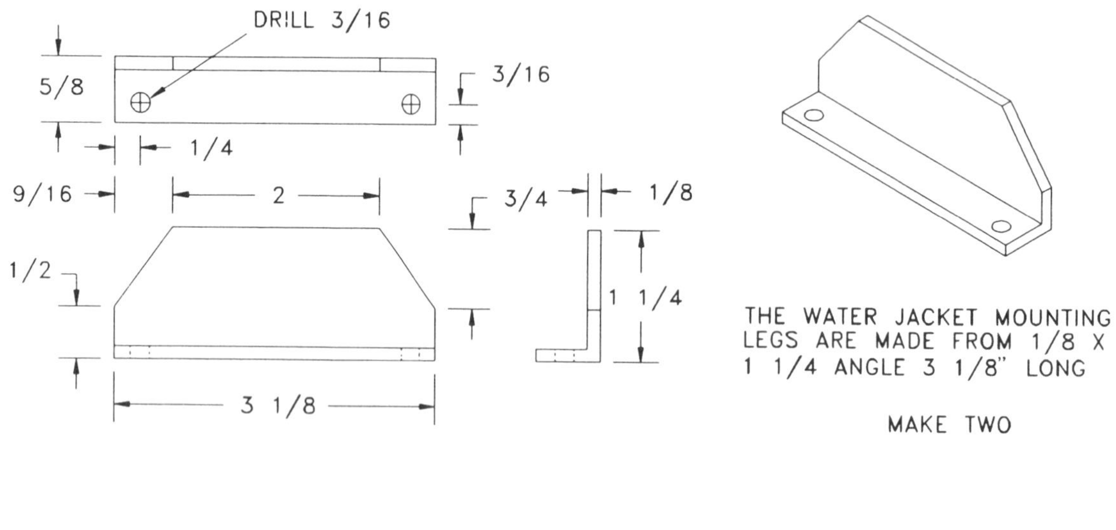

Figure 55 Water jacket legs

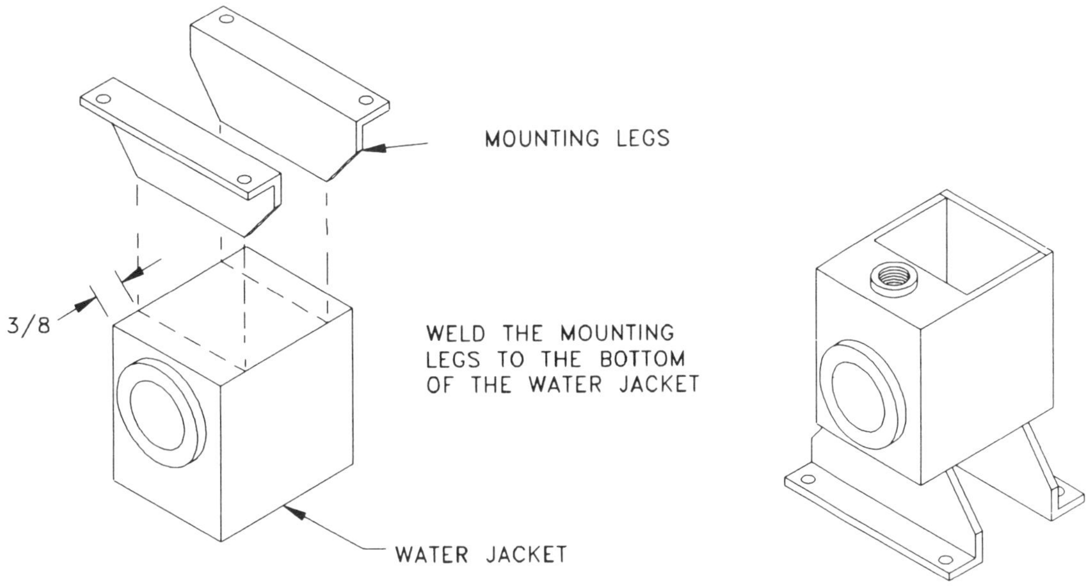

Figure 56

seriously hamper the operation of the engine. But any flaw that extends the length of the cylinder will cause problems. When the work is finished flush the cylinder thoroughly to remove all traces of abrasive.

The two mounting legs for the water jacket are shown in figure 55. Each one is made from a piece of 1/8" x 1-1/4" angle iron 3-1/8" long. Trim the base of each bracket to a width of 5/8". Mark and drill the 3/16" holes and mark and cut the angled corners. Clean up any rough edges with a

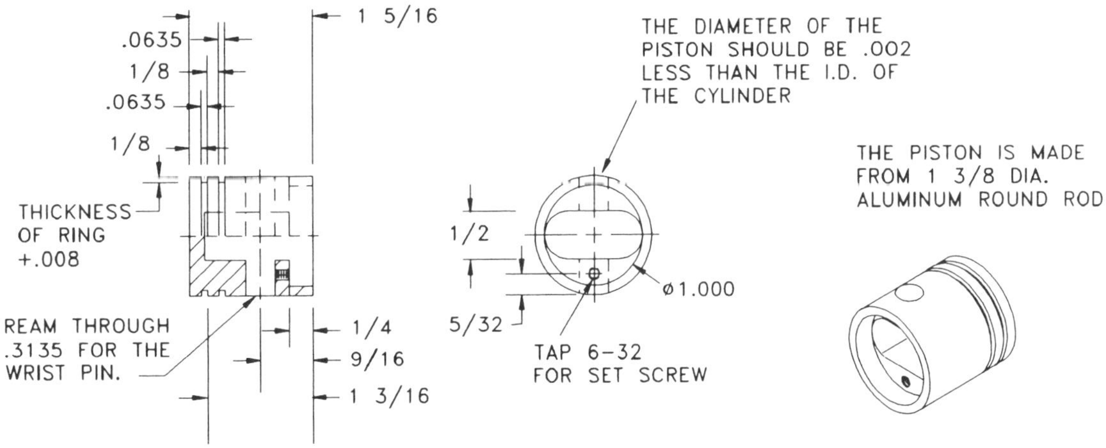

Figure 57 The piston

file.

Weld the brackets to the bottom of the water jacket as shown in figure 56. Be sure to cover the cylinder ends to protect them from weld splatter.

MAKING THE PISTON: The piston is made from a 1-3/8" diameter aluminum round rod. It is shown in figure 57. Before making the piston it is a good idea to have the rings in hand so that you know what size to make the ring grooves. We purchased our rings from Power Model Supply. They are 1-1/4" rings, 1/16" thick and their part number is RRR-102. The address and phone number for Power Model Supply is 13260 Summit drive, DeSoto, MO 63020, Phone (314) 586-6466.

To begin making the piston, chuck a length of 1-3/8" diameter aluminum round rod in a 3-jaw chuck so that 1-3/4" sticks out past the jaws of the chuck. Face off the end and turn the outside diameter of the piston to a diameter of .002" less than the diameter of the cylinder. The object is to have the piston sized so that it travels in the cylinder smoothly with minimum side play.

Cutting the grooves for the piston rings requires special care. The finished ring grooves should be about .001" wider than the piston ring and about .008" deeper than the thickness of the rings. The cut off tool to form the ring grooves should be slightly narrower than the ring and two or more passes are needed to cut the groove to finished width. The set up for cutting the ring grooves is the same as that for the cut off operation. The tool should be set at a right angle to the work in the proper position to cut the groove. Lock the

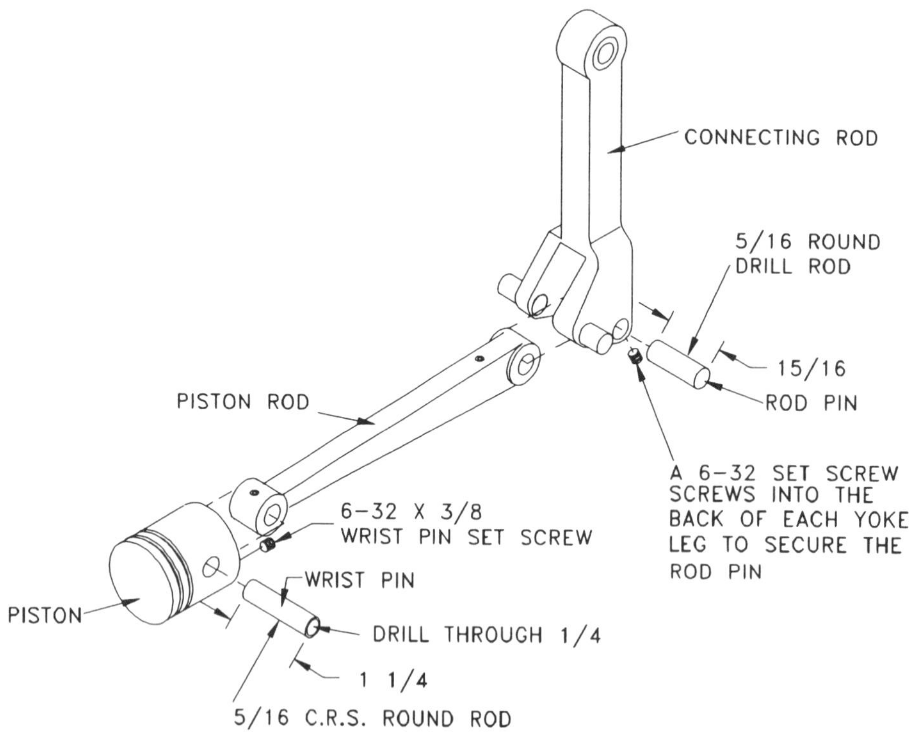

Figure 58 **Assembling the linkage**

carriage and cut the groove to a depth .008" greater that the thickness of the ring. Check the ring for fit as you proceed. The ring should fit the finished groove freely with minimum side play.

Center drill the end face and step drill a 1/2" hole 1-3/16" deep in the end of the piston for the piston rod. Counter bore the end to diameter 1", 1/4" deep. Cut off the piston 1-5/16" long. To cut the slot in the piston, clamp it in a mill vise using a V-block to hold it securely. Center the piston with the spindle of the mill and cut the slot 1" wide 1-3/16" deep with a 1/2" end mill.

Align the piston in a V-block and clamp it in position to drill and ream the .3135" wrist pin hole.

ASSEMBLING THE LINKAGE: Assemble the connecting rod, piston rod, and piston as shown in figure 58. Begin by fastening the piston to the piston rod and the piston rod to the connecting rod. Figure 58 also shows dimensions for the 5/16" x 1-1/4" wrist pin and the 5/16" x 15/16" rod pin. Secure both pins with 6-32 x 3/8 set screws as shown.

Set the pivot post on the base and position it so the drilled hole in the leg of the post aligns with the marked location in the base. Clamp it in position with a C-clamp. The final position of the pivot post will be found later by trial with the cylinder clamped in place. Slide the connecting rod onto the crank pin. Screw the 1/4-28 grease fitting with flat washer into the crank pin.

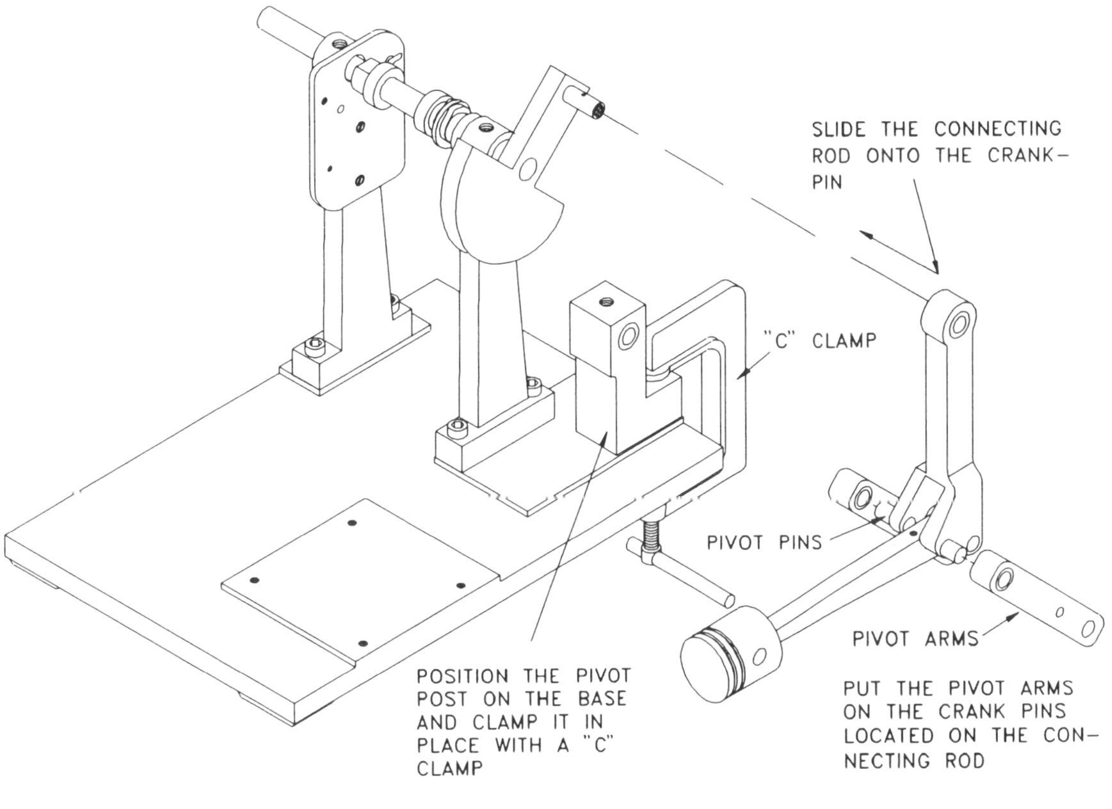

Figure 59

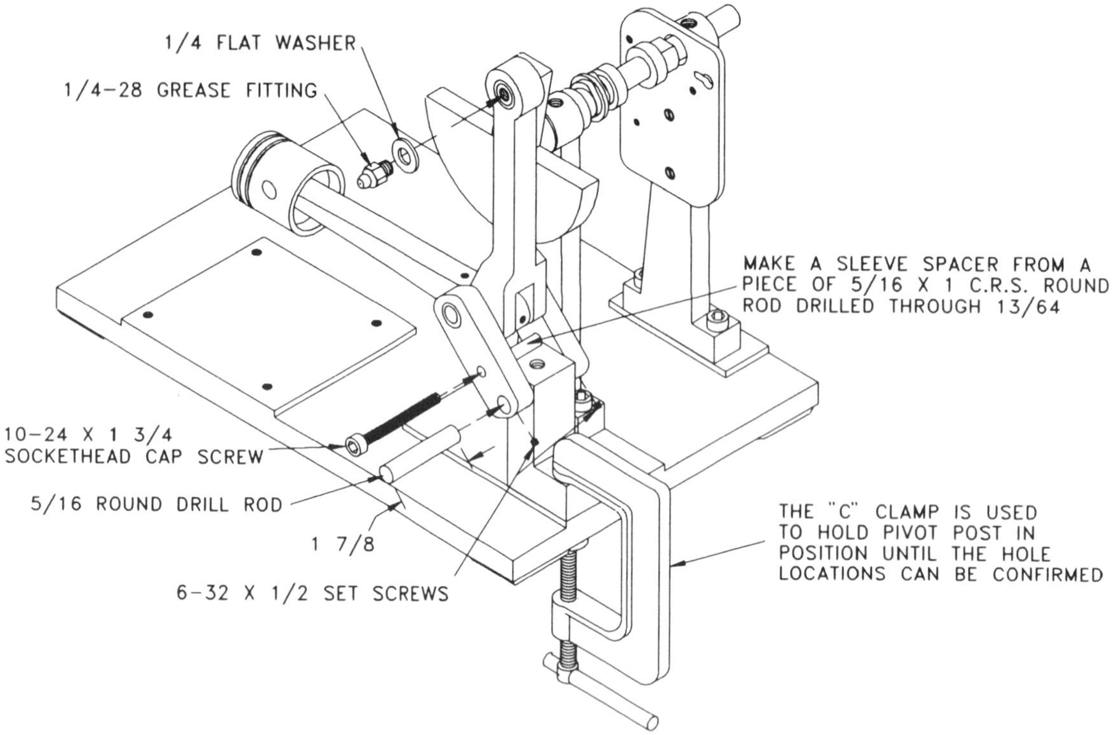

Figure 60

The purpose of the grease fitting is to lubricate the crank pin.

Put the pivot arms on the pivot pins on the connecting rod. Slide the 5/16" x 1-7/8" pivot pin through the outside pivot arm, through the pivot post and into the inside pivot arm. Lock the pivot pin in position with a 6-32 x 1/2" set screw screwed into the bottom of each pivot arm as shown in figure 60. Notice the sleeve spacer between the pivot arms in figure 60. It is a 1" length of 5/16" C.R.S. round rod drilled through 13/64". Slide the 10-24 x 1-1/4" socket head cap screw through the outer pivot arm and through the sleeve spacer and thread it into the inside pivot arm.

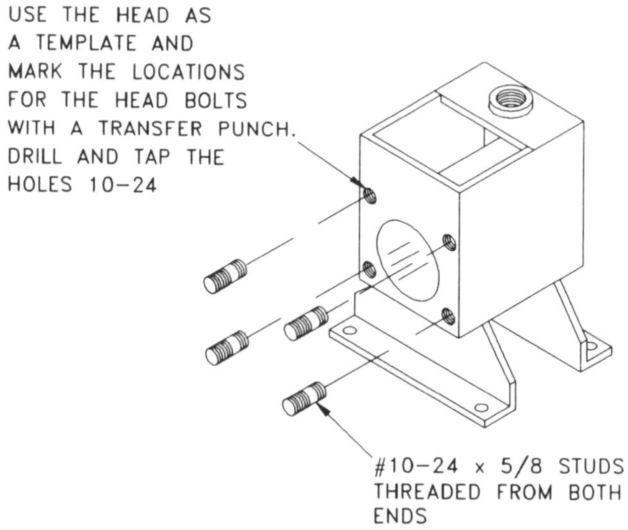

Figure 61

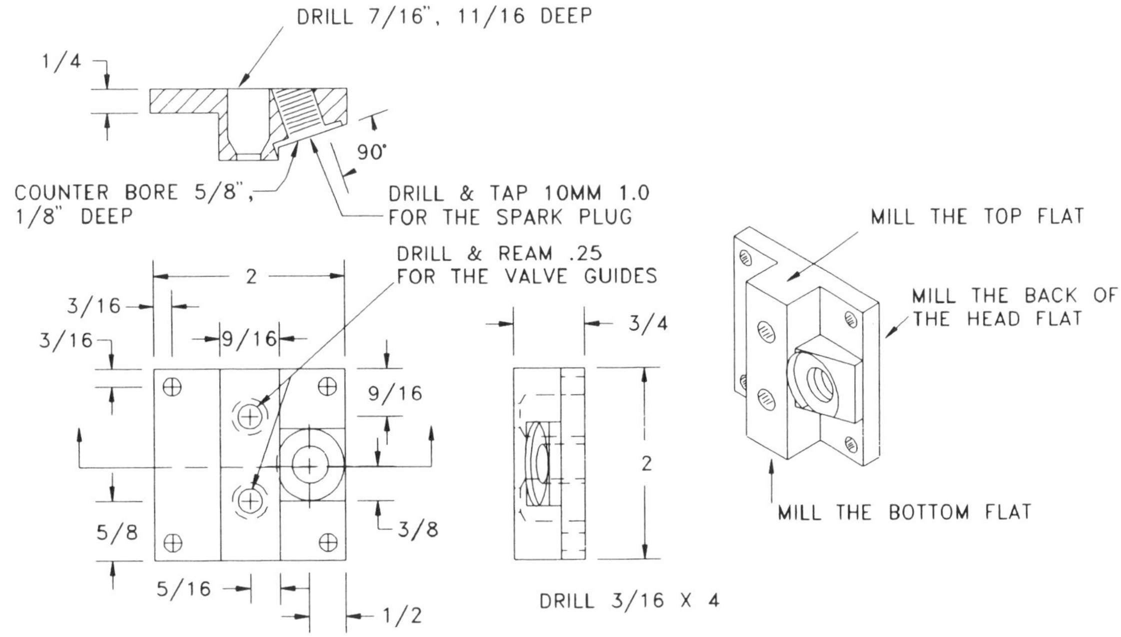

Figure 62 The head

THE HEAD: The locations for the holes in the head are shown in figure 62. Begin by chucking and centering the head in a 4-jaw chuck and face off the back side. Mill all four sides of the head, the finished size to be 2" x 2". Locate and drill the four 3/16" mounting holes. Place the head in position on the end of the water jacket. Use a 3/16"

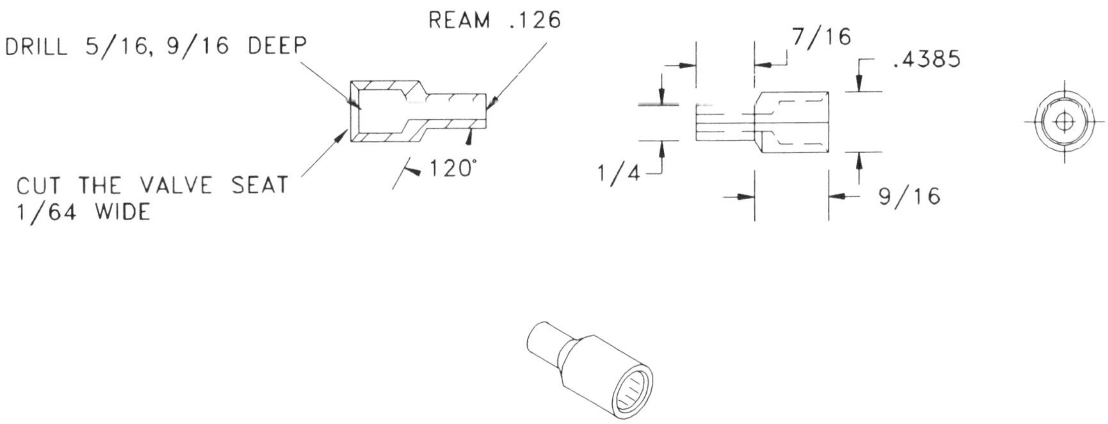

Figure 63 Valve guide. Make two

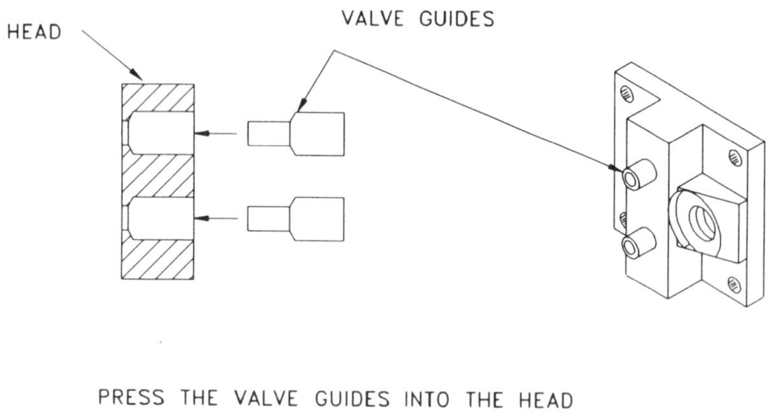

Figure 64

damage the cylinder walls. Remove the head and lay out the hole locations for the valve guides. Check to be sure that the holes will fit inside the scribed circle or the valve heads could jam against the end of the water jacket when they open. It is important to drill the valve guide holes in the head straight for proper valve action. Drill both the .25" and the 7/16" holes from the cylinder side of the head. Drill and ream through .25" for the first valve guide. Increase the size of the hole to 7/16", 11/16" deep. It would be easy to drill the 7/16" hole too deep so use a drill stop set at

transfer punch to mark the hole locations on the water jacket. Drill and tap the holes in the water jacket 10-24. Make four 5/8" long 10-24 studs threaded on both ends. Thread one end back 1/8" and thread the other end back 3/8". This leaves an unthreaded section in the middle for the studs to tighten down against. Apply a little thread lock to the ends threaded back 1/8" and screw them into the water jacket. See figure 61. Secure the head to the water jacket with 10-24 nuts. Mark the location of the cylinder opening on the back side or cylinder side of the head. This is done with a scribe from the inside of the cylinder so be careful not the

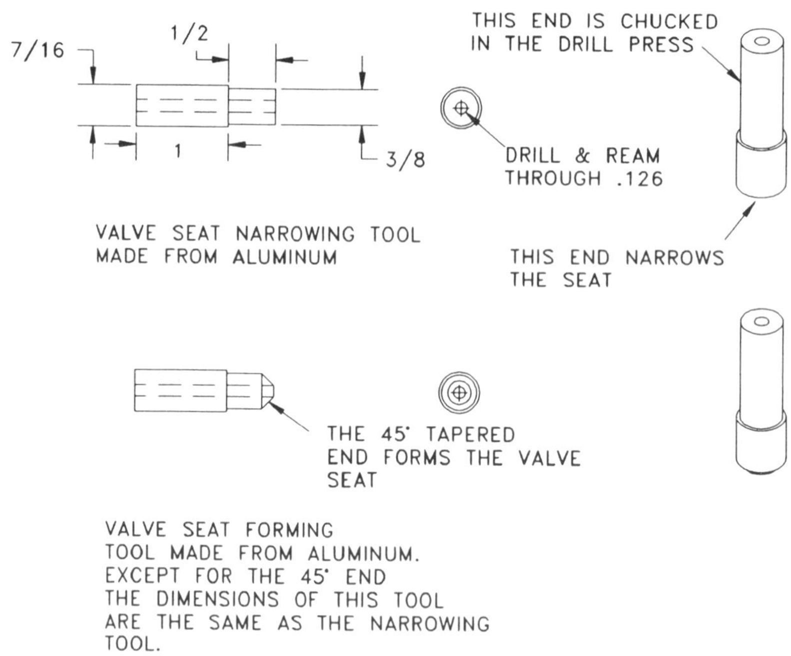

Figure 65 The valve seat forming tools

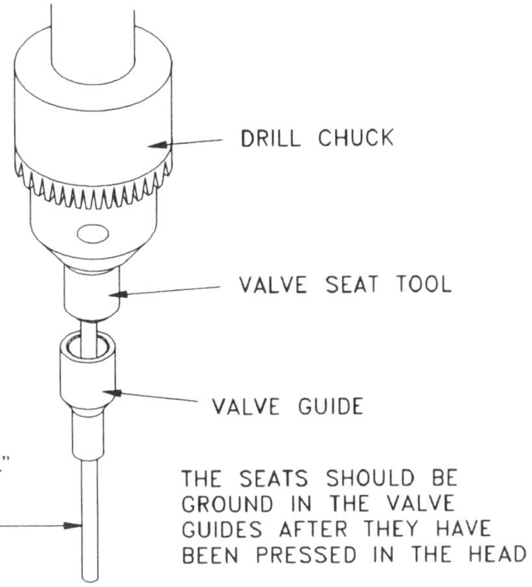

Figure 66 **The valve grinding set up**

a drill stop set at the 11/16" depth as a precaution to prevent drilling all the way through the head. Repeat the above procedure to drill the hole in the head for the other valve guide

Drill and tap the spark plug hole 10MM 1.0 at 90 degrees in reference with the sloped surface at its location. Counter bore the spark plug hole with a 5/8" milling cutter 1/8" deep.

MAKING THE VALVES AND VALVE GUIDES: The valve guides are shown in figure 63. You will need a 1/4" collet to make the valve guides. For review information on making your own collets refer back to figure 22 on page 27.

Begin making the valve guide by chucking a piece of 1/2" C.R.S. round rod in a collet so that 1-1/4" sticks out past the chuck. Turn the outside diameter to 1/4" 7/16" back from the end. Set the compound rest and turn the 60 degree tapered transition. Part off at 1". Mount the blank in a 1/4" collet. Center drill and drill and ream through .126". Increase the size of the hole to 5/16", 9/16" deep.

Coat the outside of each valve guide with epoxy and press them into the head as shown in figure 64. The reason for the epoxy is to prevent possible air leaks around the guides that could affect the engine compression. The seat end of the valve guide should set about 1/32" below the surface on the cylinder side of the head and the neck end will extend out the outside of the head about 1/4".

The valve and stem are shown in figure 68. The stems are made from a piece of 1/8" drill rod 1-5/8" long. The valve heads are made from a piece of 7/16" C.R.S. round rod. To make the valve heads, chuck a piece of 7/16" round rod in a collet. Face off, drill and ream through .126". Turn a .190" shoulder on the end 1/8" back from the end. Adjust the compound rest and turn the 45 degree bevel. Cut off the valve head long enough to allow for facing off. Repeat to make the second valve head.

Cut two pieces of 1/8" diameter drill rod to 1-5/8" in length for the valve stems. Use a V-block to hold the stem in position and drill a 1/16" hole in the end of each one, 1/16" from the end.

Insert a stem in the 1/8" hole in a valve head so that the stem is even with the top of

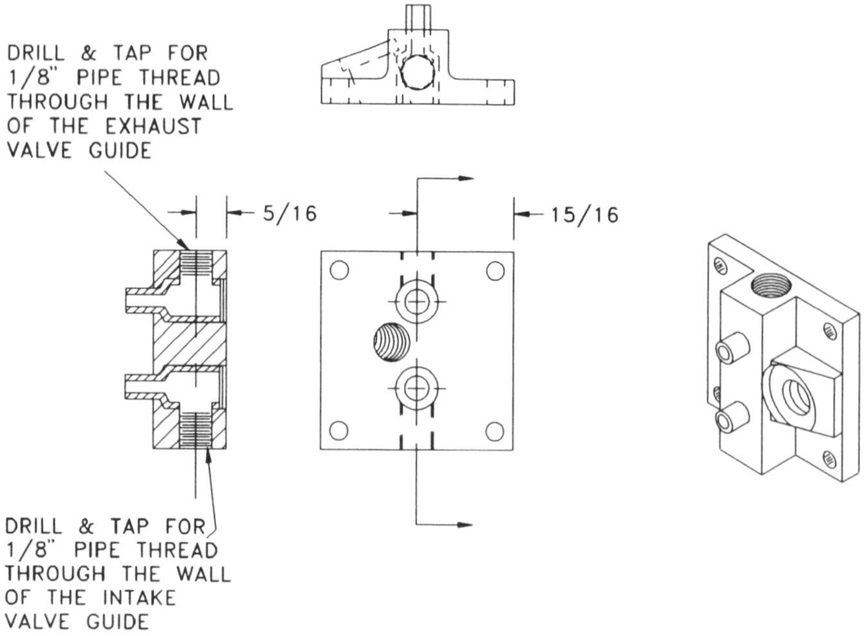

Figure 67

for the second valve.

Mount the valve stems in an 1/8" collet and face off the valve heads and true up the 45 degree taper. The finished thickness of the .380" diameter section of the head to be .070".

In a commercial operation the shop will have a valve grinding tool with an assortment of stones to grind the valve seats. We are not so fortunate and are faced with making our own tool. The idea for this one came from Jim Lewis and I must say it works very well. The principle is the same as that used to lap the cylinder. The valve seat tool and seat narrowing tool are made of aluminum and then charged with lapping compound. The seat tools are shown in figure 64 and the setup is shown in figure 65. To form the seat, charge the end of the seat tool with valve grinding compound and chuck it in a drill press. Insert the pilot shaft through the valve guide and into the valve seat forming tool. The purpose of the pilot shaft is to keep the valve guide in perfect register with the tool. While running the drill press at a slow to moderate speed force the valve guide against the seat tool giving it a 1/4 turn every minute or so until the width of the seat measures about 1/64". During the process be sure and hold the end of the pilot shaft to keep it from spinning in the valve guide and enlarging the guide hole. If you over do it and grind the seat too wide you can narrow it with the narrowing tool which has a flat end.

When the seat has been formed, lap the valves with valve grinding compound. One way to do the lapping is to insert the valve into the valve guide and hold its stem with a pin vise. Apply fine grit lapping compound and twist the valve back and forth against its seat rotating a 1/4 turn every fourth or fifth twist. You can tell when the valve has been lapped when upon close examination you can see a narrow, uniform gray line around the surface of the seat and

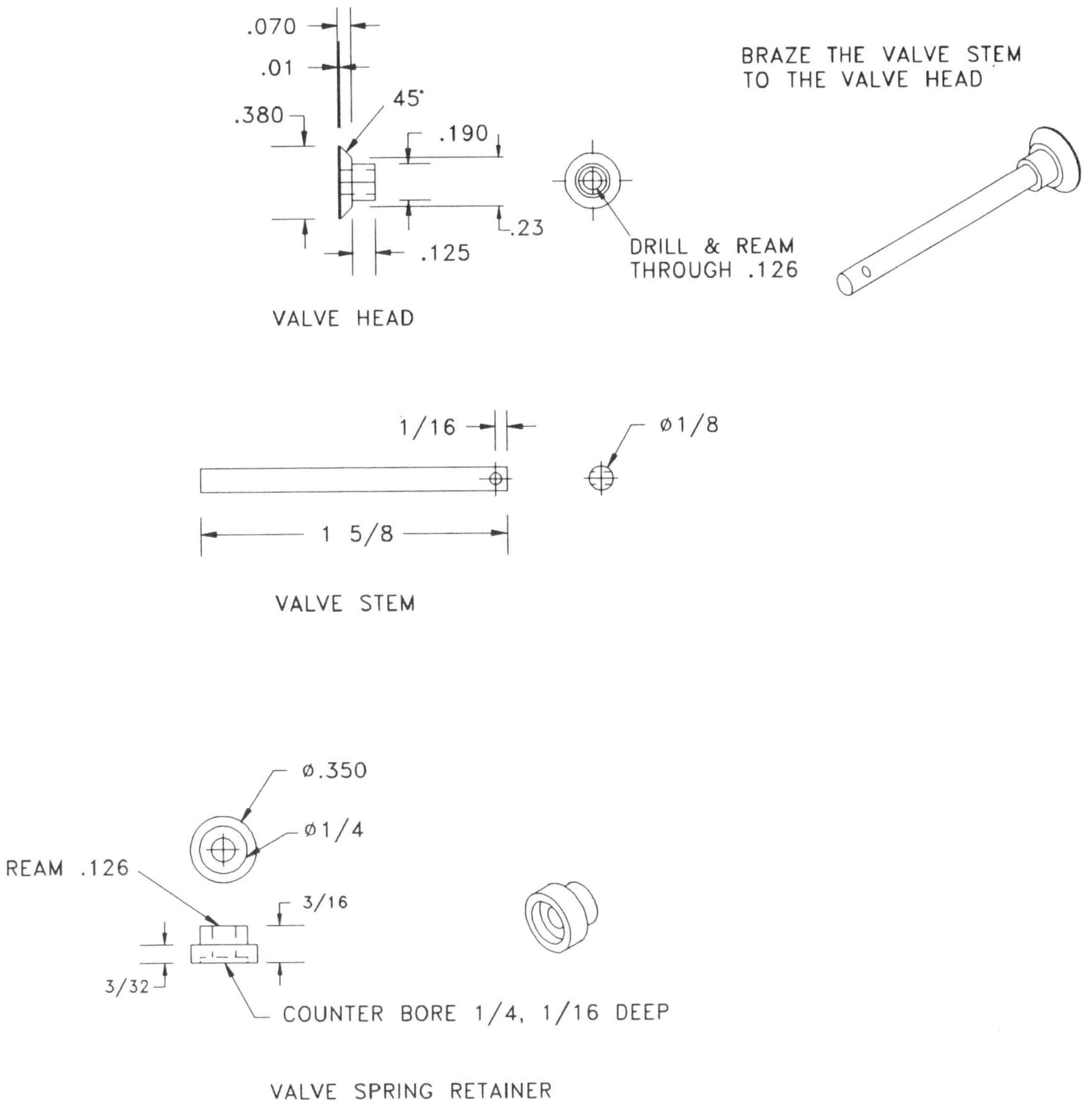

Figure 68 The valves and retainer

the valve. Be sure and mark the valves in such a way that you will know which seat they are made for.

Drill through the top of the head and through the wall of the valve guide where shown in figure 67. Tap for 1/8" pipe thread. This is the exhaust port. Drill through the bottom of the head and through the wall of the intake valve guide where shown. Tap for 1/8" pipe thread. This is the intake port.

The valve retainers are shown in figure 68. To make them, chuck a piece of 3/8" diameter C.R.S. round rod in 3 jaw chuck. Turn a 1/4" diameter shoulder 3/32" from the end. Part off at 3/16" from the end.

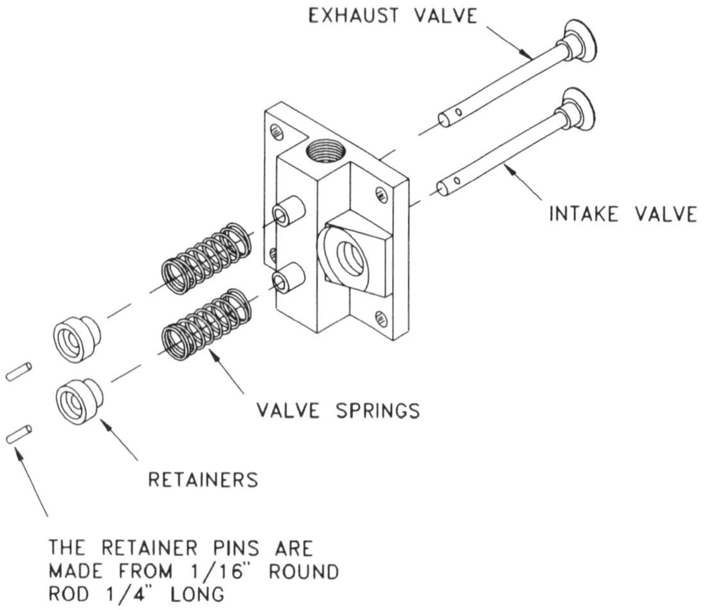

Figure 69 Installing the valves

Mount the 1/4" diameter shoulder in a chuck or collet and counter bore 1/4", 1/16" deep.

Figure 69 contains an exploded view of the valve assembly. The valve springs we used were purchased from the local hardware store. They were made by the Gardner Spring Co. located in Tulsa Ok.. Part #C14C. The dimensions are: .312" OD x .0286" wire size x 1" long. Several manufactures make these springs and they can be found at most hardware stores. A lot of people make their own springs out of music wire so that's another possibility.
Insert the retainers into the top of the springs and put the springs on the head so they slide over the valve guides. Insert the valves and compress the springs so that the 1/16" hole in the valve stem extends out far enough to put the retainer pin in the hole. When the spring is decompressed, the pin will rest inside the counter bore in the retainer and hold the valve assembly together.

ALIGNING THE LINKAGE: Set the water jacket on the base and align the holes in the legs as close as possible with the marked locations on the base and clamp it in place with a C-clamp.

The goal in setting up the linkage, is to be able to turn the crank a complete revolution without binding or jamming against other parts. The water jacket needs to be set in position so that there will be a .020" clearance between the head of the piston and the front end of the cylinder at the top of the exhaust stroke. Figure 72 demonstrates this.

We have built 4 engines and the linkage travel pattern was a little different on each one. Chances are the marked hole locations in the base are going to be very close to where they belong, but slight differences in hole locations, heights on the pillars, pivot post and water jacket as well as linkage lengths drastically affect the travel pattern of the linkage. Temporarily clamping the water jacket and pivot post in position gives us a chance to adjust their locations as needed. To begin the adjustment process, slowly turn the flywheel through one complete revolution. The two problems that you are most likely to encounter are the piston rod hitting either the top or the bottom of the back edge of the cylinder and the pivot link spacer hitting the front of the pivot post. Since it is almost impossible to determine the position of the pivot post and water jacket we have to fine tune their locations by trial and error. That is, you

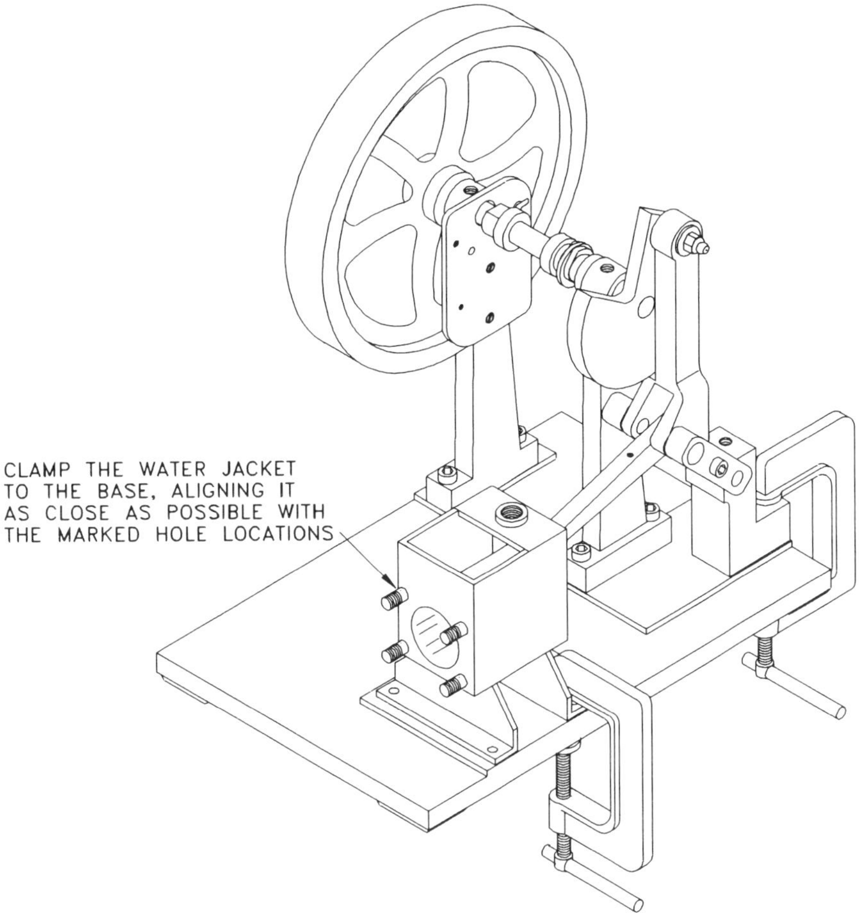

Figure 70

move the pivot post or water jacket forward or backward slightly and then rotate the flywheel to see what differences the changes made. On one occasion we had to put a 1/16" spacer under the pivot post. On another occasion the hole in the crank for the connecting rod pin was off a little and we were forced to remake it. Remember, slight location differences as little a 1/32" make a drastic difference in the travel pattern of the linkage. The best way to work through the adjustment process is to relax and pull up a chair. Slowly

Figure 71

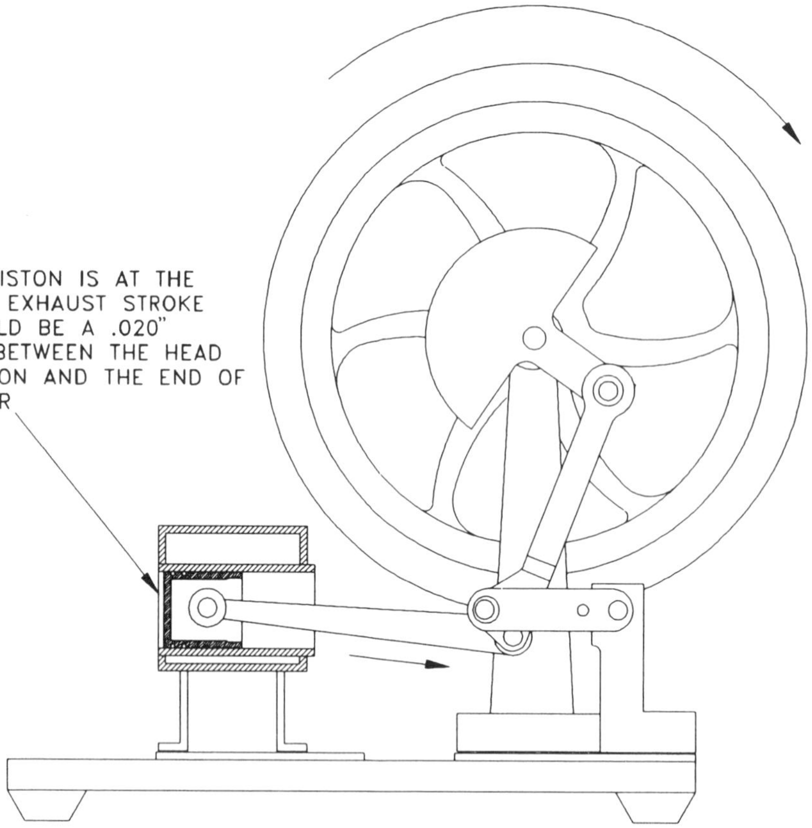

Figure 72

turn the crankshaft through a complete revolution paying careful attention to the travel pattern of the linkage. If necessary, Make careful adjustments to the pivot post and or water jacket taking note of each change and then checking the result of your changes by running the crankshaft through a revolution. With patience and careful study you will have the pivot post and water jacket positions located within a matter of minutes. When you have done so, mark the hole locations on the base for the water jacket carefully with a transfer punch. Drill and tap them 10-24. Bolt the water jacket to the base with 10-24 x 3/8" socket head cap screws.

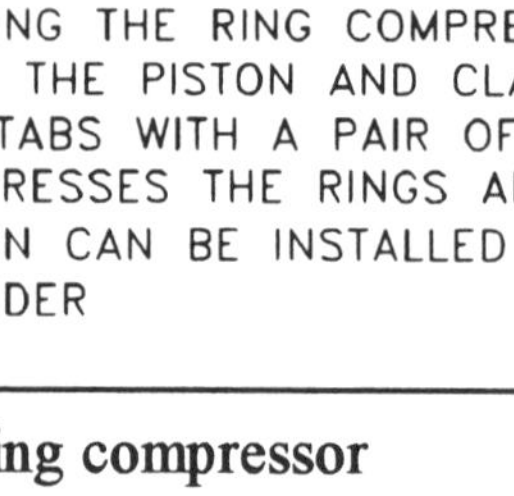

Figure 73 Ring compressor

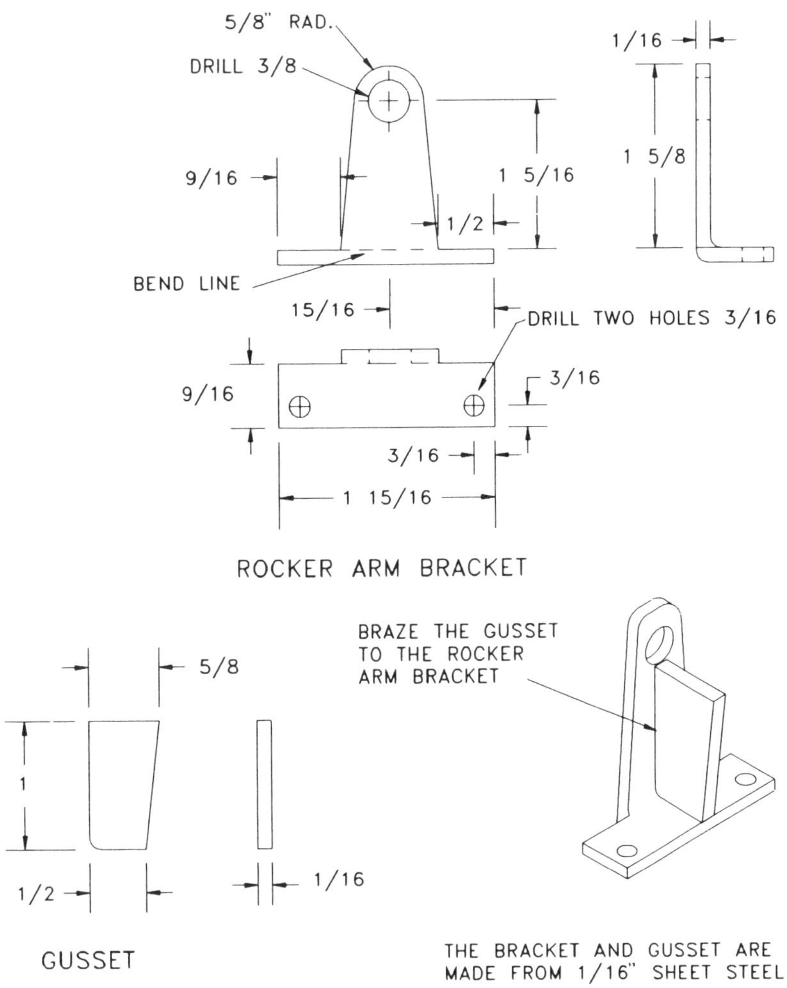

Figure 74 The rocker arm bracket

Mark the location and drill the holes in the base for the pivot post. Mark the 1/4" hole with a transfer punch. You can find the second hole in the base for the pivot post by measuring 3/4" back from the center of the 1/4" hole. Drill the hole 5/16". Bolt the pivot post to the base with a 1/4-20 x 1/2" bolt, nut, and flat washer and a 5/16-18 x 1" bolt with flat washer.

INSTALLING THE RINGS: Before installing the rings on the piston, try them in the cylinder to ensure at least a .004" clearance at the slit. Place the rings on the piston. It is necessary to compress the rings so the piston can slide into the cylinder. You can make a ring compressor out of light gauge sheet metal. A drawing of one is shown in figure 73. It's simply a 1" length of sheet metal formed to the diameter of the piston. A 1/2" tab is bent out on each end. When installing the piston the tabs are clamped with a pair of pliers. This compresses the rings into the ring grooves. The back of the piston is tapped lightly with a wooden handle and the piston slides into the cylinder. Don't force the piston, it should slide in smoothly.

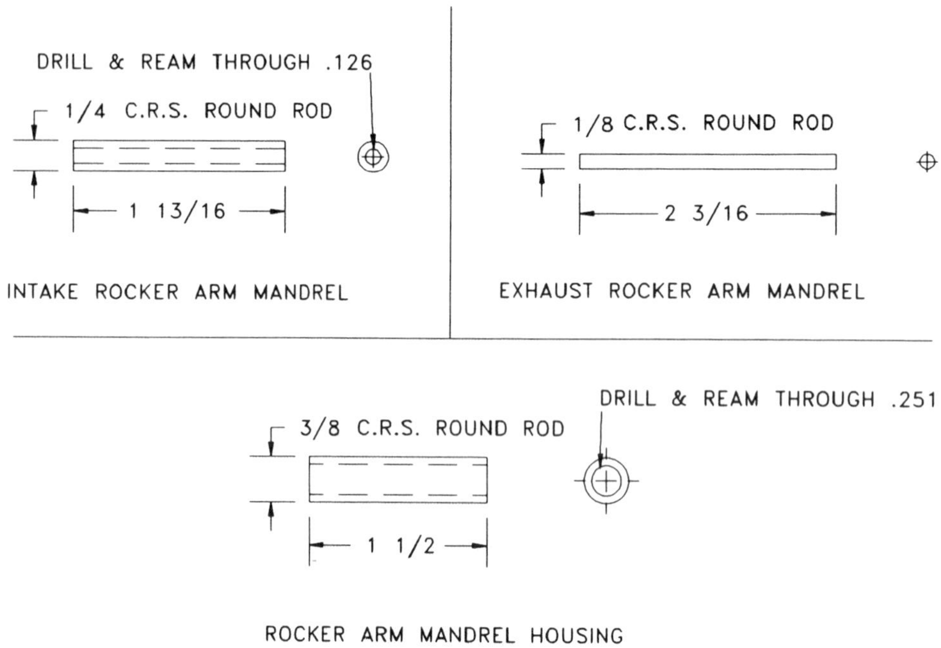

Figure 75

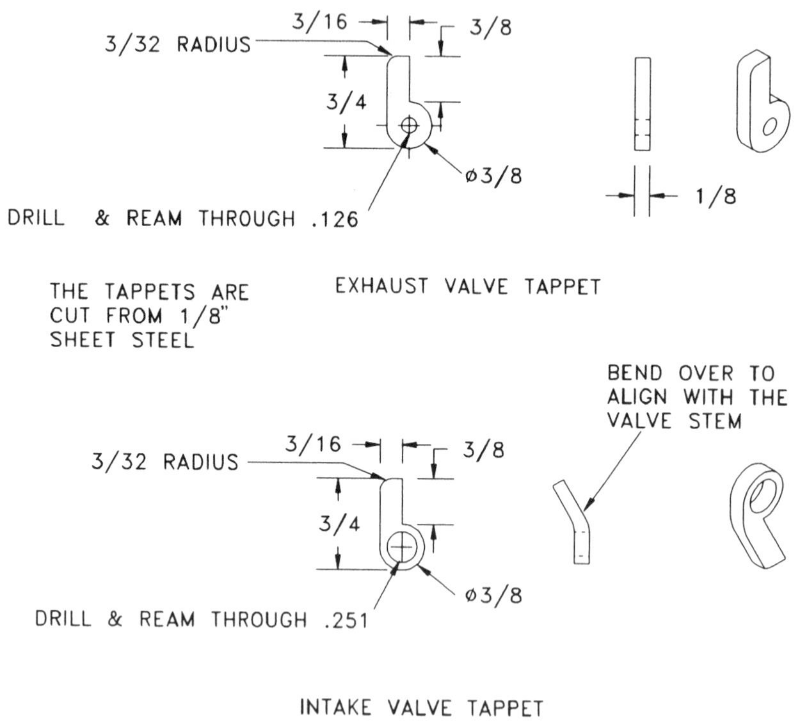

Figure 76

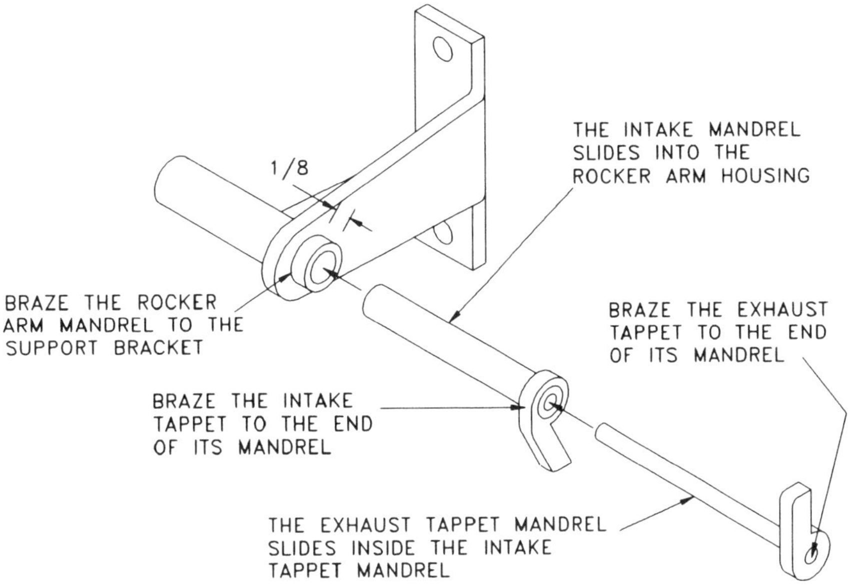

Figure 77 The rocker arm assembly

THE ROCKER ARM ASSEMBLY: Figure 77 shows an exploded view of the rocker arm assembly. Begin by making the intake and exhaust mandrels and housing. They are shown in figure 75. The mandrel housing is made from 3/8" C.R.S. round rod 1-1/2" long. Drill and ream through the mandrel housing .251". The intake mandrel is made from 1/4" C.R.S. round rod 1-13/16" long. Drill and ream through it .126". The exhaust mandrel is made from 1/8" C.R.S. round rod and it is 2-3/16" long.

The simple layout for the rocker arm bracket is shown in figure 74. Lay the pattern out on a flat piece of 1/16" sheet steel. Cut the shape out on a bandsaw then drill the two 3/16" holes and the 3/8" hole. The 90 degree bend can be made in a vise. Make the gusset and braze it in place as shown.

The tappets are shown in figure 76. They are made from 1/8" thick steel. They are simple to make and there really isn't much explanation necessary. Make them to near shape and then finish to size with a file. The bend in the intake tappet will be made later so that it will align with the valve stem.

Put the rocker arm assembly together as shown in figure 77. Braze the mandrel housing to the bracket and braze the intake tappet to the end of its mandrel and the exhaust tappet to the end of its mandrel. The intake mandrel then slides inside the mandrel housing and the exhaust mandrel slides inside the intake mandrel

Bolt the head and rocker arm assembly to the water jacket with four #10-24 nuts as shown in figure 78. Make the head gasket out of .020" thick paper gasket material.

MAKING THE PUSH ROD LINKAGE: The push rods are what open and close the valves by way of cam action. There are two push rods, one to open and close the intake valve and the other to open and close the exhaust valve. The cam side of each push

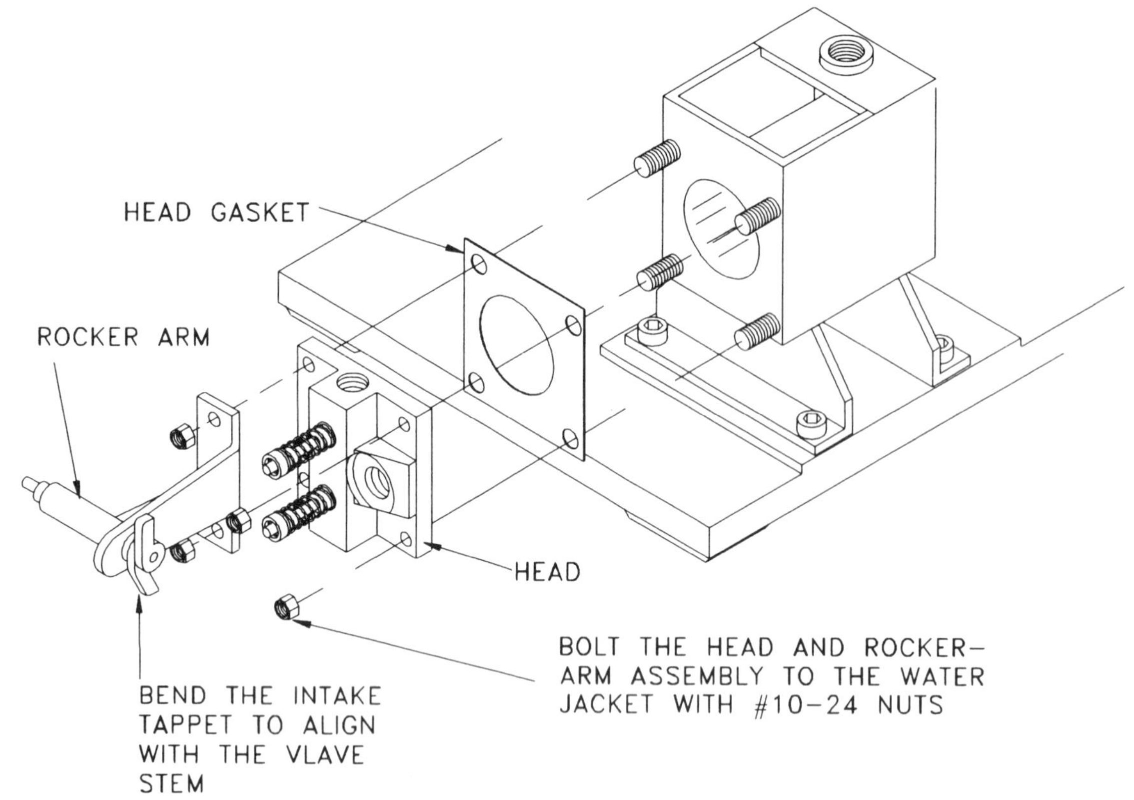

Figure 78 **Bolting the head and rocker arm assembly to the water jacket**

rod has a yoke end brazed to it. The yoke slips over the crankshaft and rests between the cams. A roller referred to as the cam follower, is mounted on the yoke. In operation the lobe of the cam comes around and hits the roller and forces the valve open. When the lobe rotates off the roller the valve closes.

The rocker arm side of the push rod is attached to the rocker arm mandrel with a link that hinges on the end of the push rod and clamps to the end of the mandrel.

Make the push rods out of 3/16" key stock. See figure 79. The length of the exhaust push rod is 8-5/8" and the length of the intake push rod is 8-7/8". Cut a 1/8" notch 1/2" back on one end of each push rod. Drill a 1/8" hole 1/8" back from each end of each push rod. Cut two 1/8" x 5/16" long C.R.S. round rod pins and braze one into the notched end of each push rod. Cut two 1/8" x 1/2" C.R.S. round rod pins. Drill a 1/16" hole 1/16" from the end of each one for a cotter pin and then braze them into the 1/8" hole in the rocker arm side of each push rod as shown in figure 81.

The yoke and cam roller are shown in figure 84. Make the yoke from 1/16" sheet steel. Cut to near size and then clean up the

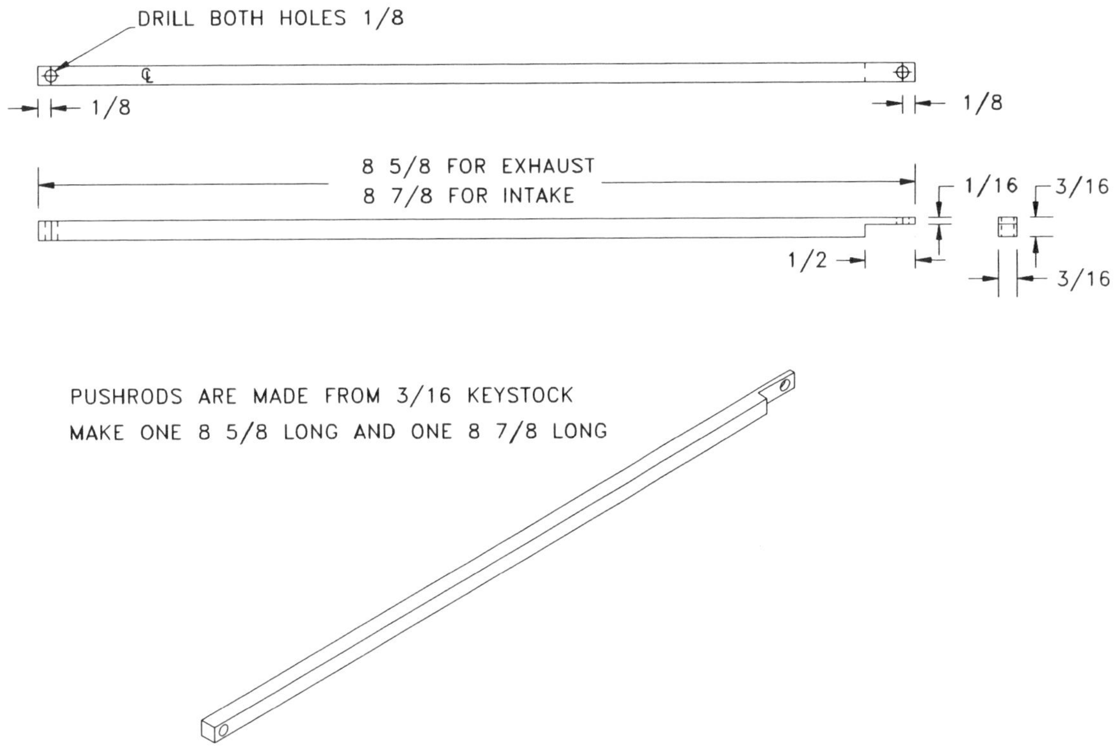

Figure 79 The push rods

edges with a file. Make the cam rollers from 1/2" diameter C.R.S. round rod. Turn the end of the 1/2" rod to .485", 1/2" back. Drill and ream .126", 1/2" deep. Cut off two rollers each .120" long.

The push rod connecting links are shown in figure 82 and 83. They're really kind of fun to make. You will need a short length of 3/16" x 1/4" C.R.S. flat bar for the exhaust link and a short length of 3/16" x 3/8" C.R.S. flat bar for the intake link. Since the links

Figure 80

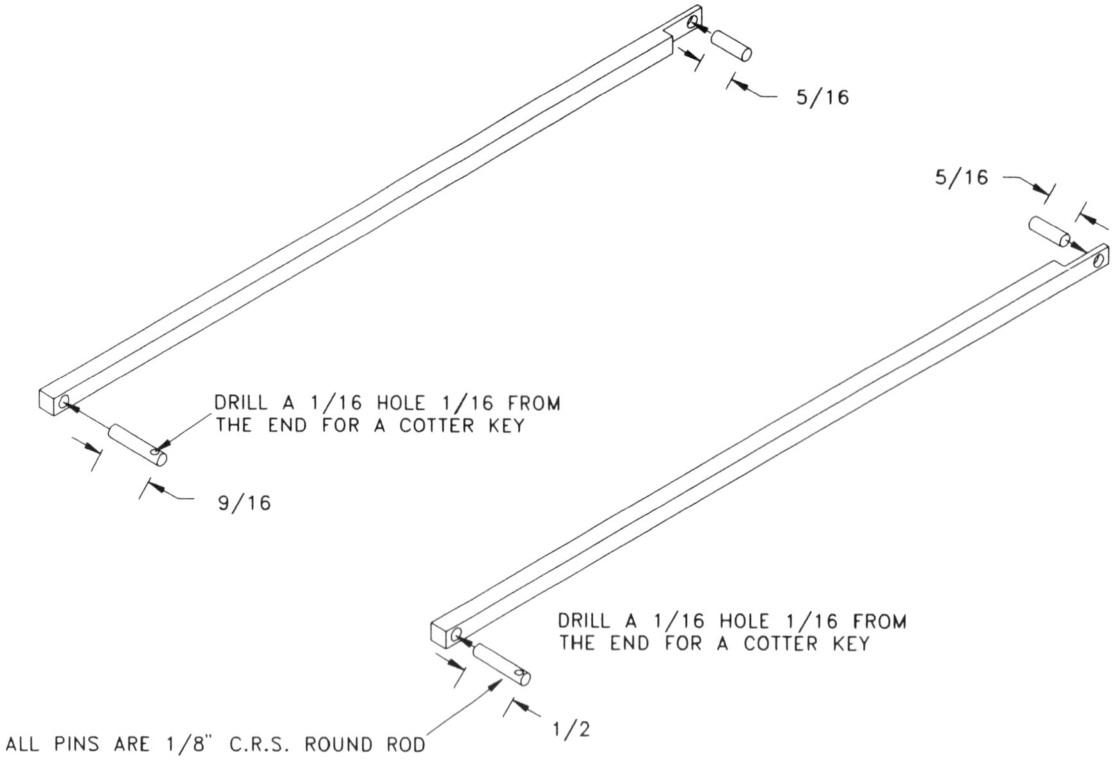

Figure 81

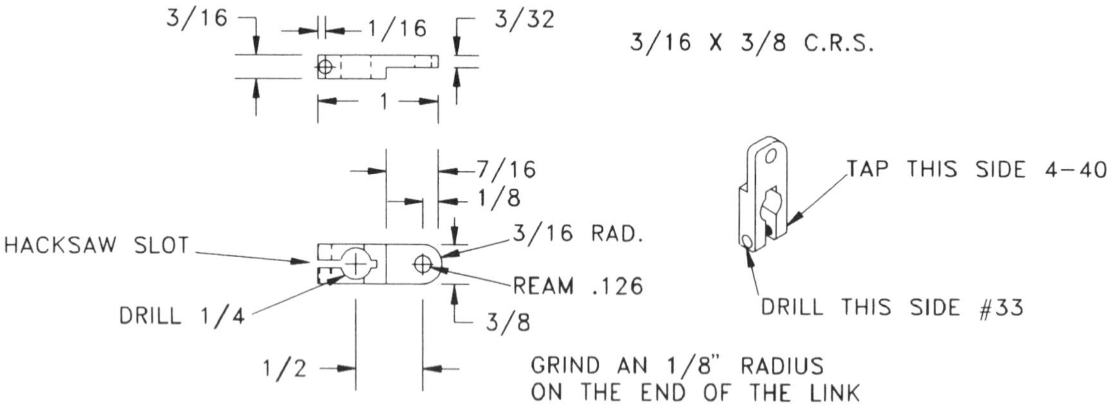

Figure 82 **Intake connecting link**

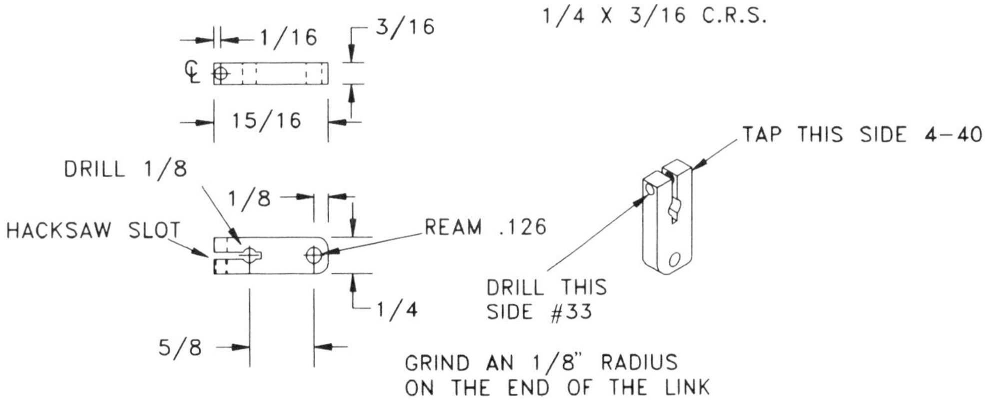

Figure 83 **The exhaust connecting link**

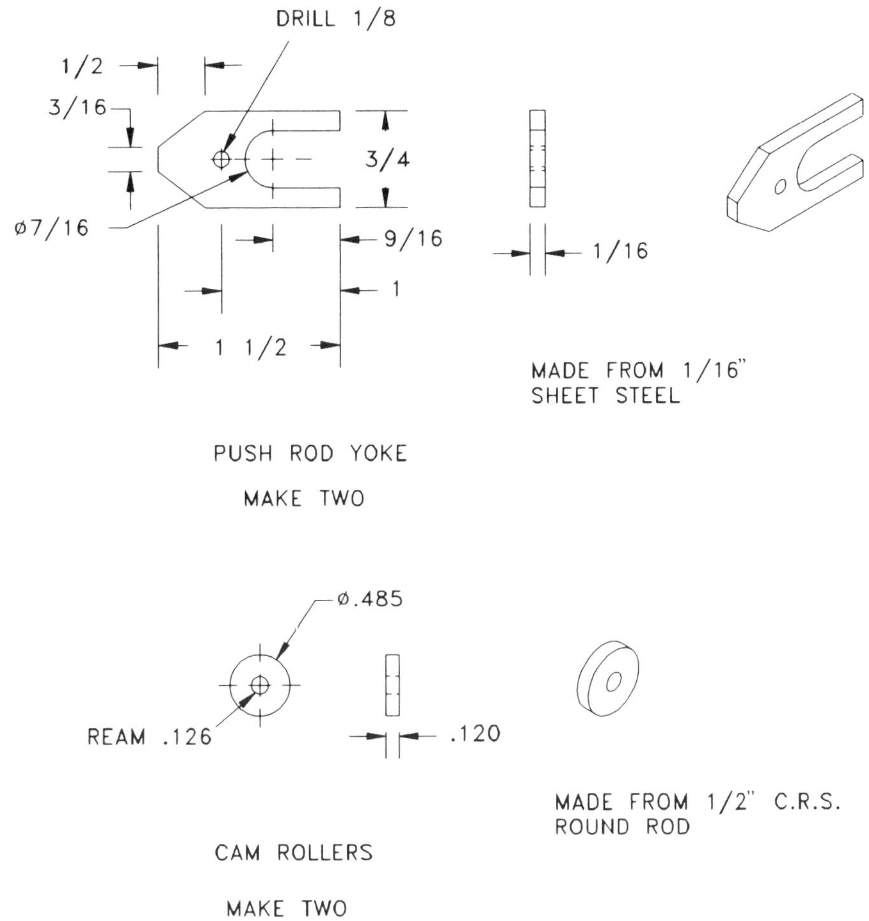

Figure 84 **The yoke and cam roller**

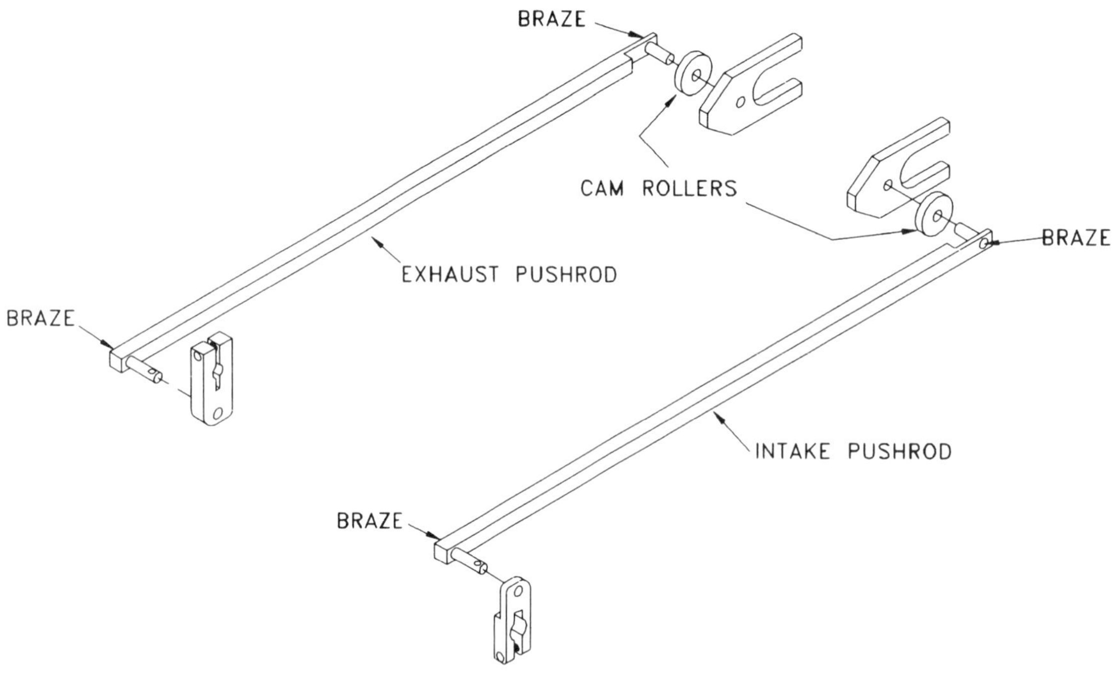

Figure 85

are such small parts, it is best to make them on the end of the stock material before cutting them to length. This gives a handle to hang on to while drilling the holes. The dimensions for the links are shown in figures 82 and 83.

Lay out the intake link on the end of the 3/16" x 1/4" blank. Drill the 1/4" hole. Drill and ream the .126" hole. Cross drill the end hole through with a #44 drill. Increase the size of the clearance hole with a #33 drill. Use the clearance hole to align the tap and tap the 4-40 hole. The step section can be cut on a milling machine or cut out with a hacksaw. Grind the radius on the end and clean up the rough edges with a file. Cut the link off at 1". Saw the slot through to the 1/4" hole with a hacksaw. The slot end of the link acts like a clamp. When the 4-40 screw is tightened, the link will clamp down on the intake mandrel.

The dimensions of the exhaust link are different than those of the intake link, but methods used to make it are the same. There is no notch in this link and the hole for the mandrel is 1/8" instead of 1/4".

Figure 85 and 86 show exploded views of the push rod assembly. Simply slide the cam rollers on the pins as shown. They should rest inside the notched area of the pushrod and have plenty of clearance and roll freely. Place the yoke on the roller pin next. Align the yoke so that it is straight and braze it to the push rod where shown.

Put the connecting links on the other end of the pushrods. They slide on the pins and are held in position by 1/16" x 1/2"

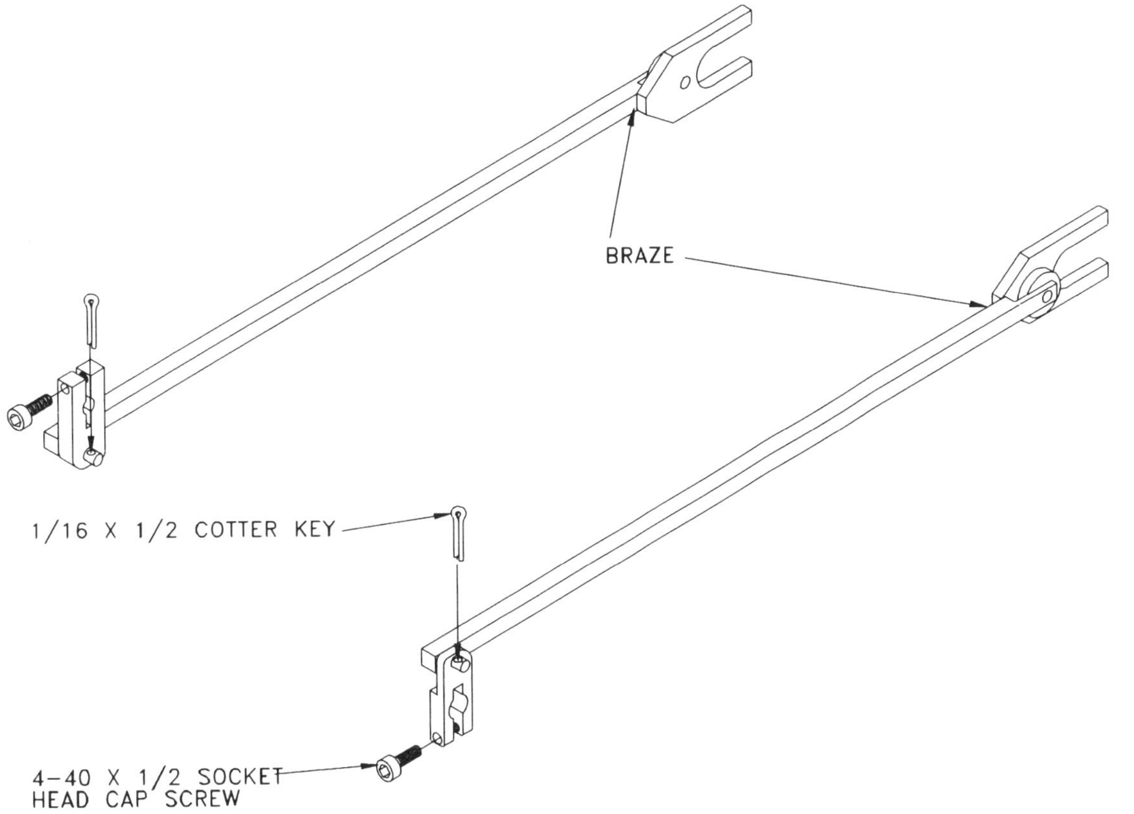

Figure 86

cotter keys. Screw the #4-40 socket head cap screws into the links at this time, but do not tighten them. Slip the yoke end of the intake pushrod onto the crankshaft in position between the cams and to the right of the spacer washer. Align the cam roller so that it will ride on the intake cam. Slide the connecting link on the other end of the pushrod onto the 1/4" diameter intake rocker arm mandrel. Position the rod so that the cam roller rests against the cam. The proper tappet clearance is about .005". After checking the tappet clearance, tighten the 4-40 screw

Figure 87

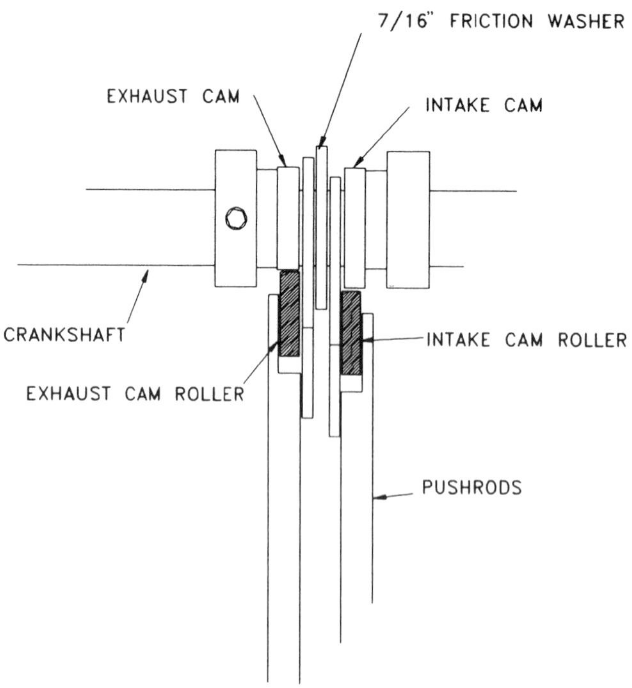

Figure 88

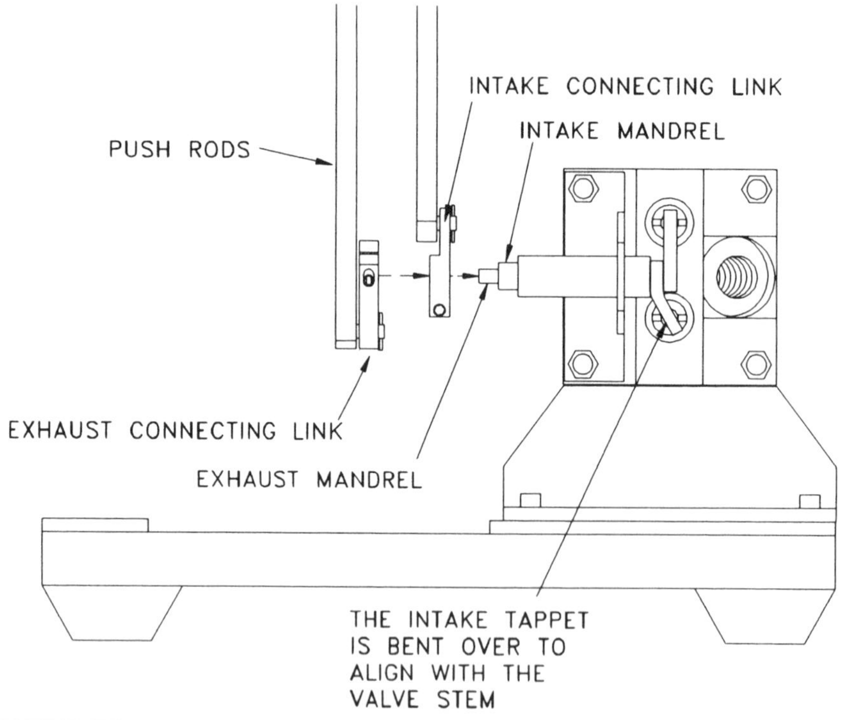

Figure 89

to clamp the link to the mandrel.

The exhaust push rod is installed like the intake pushrod except it goes on the opposite side of the spacer washer and the cam roller is set to rest on the exhaust cam. The connecting link slides on the 1/8" exhaust mandrel and the 4-40 screw is tightened to clamp the link on the mandrel. The photo in figure 80 shows the links installed. Also see figure 86 and 89. The photo in figure 87 and the drawing in figure 88 show how the yoke end of the pushrods look installed.

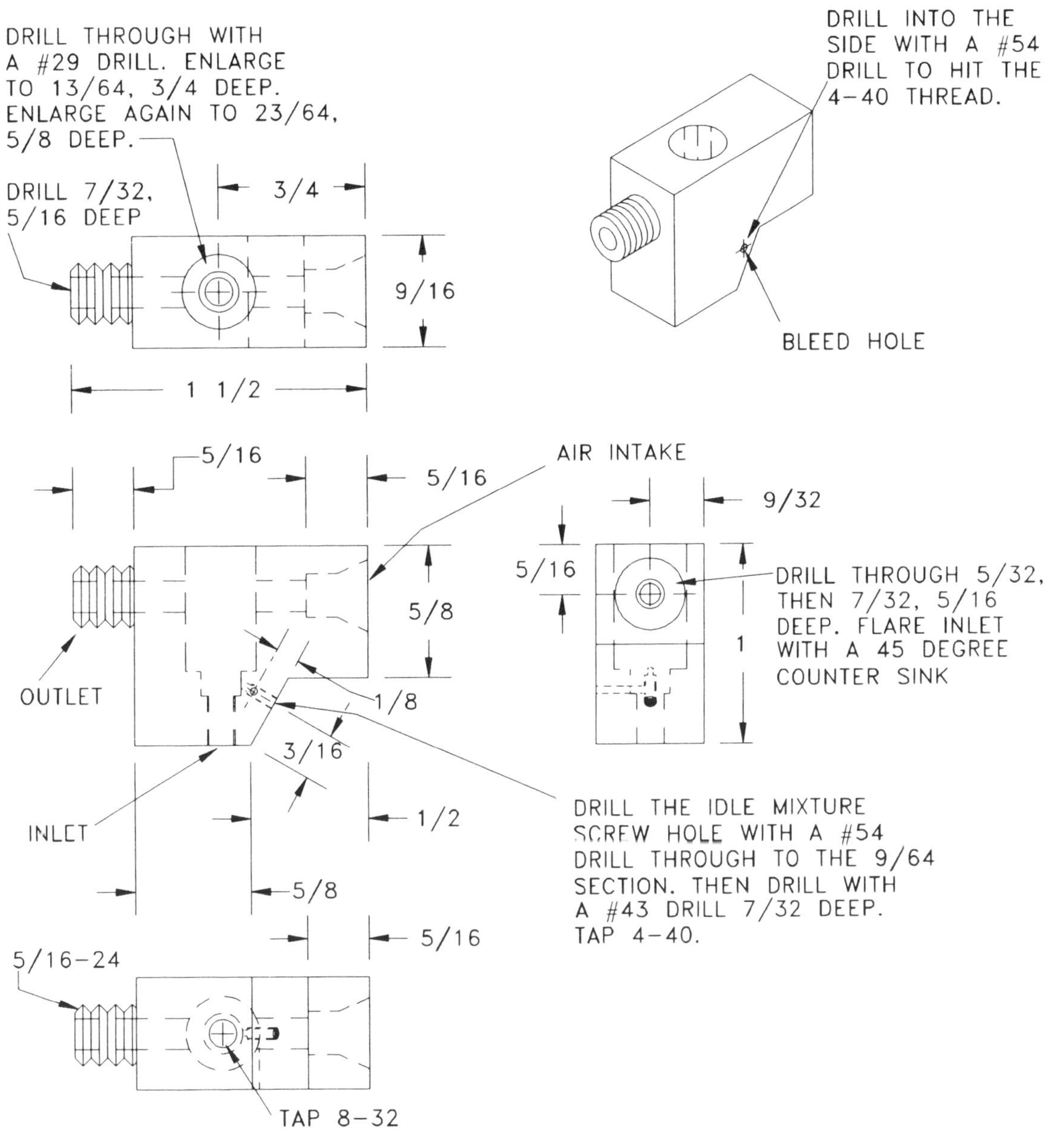

Figure 90 The carburetor body

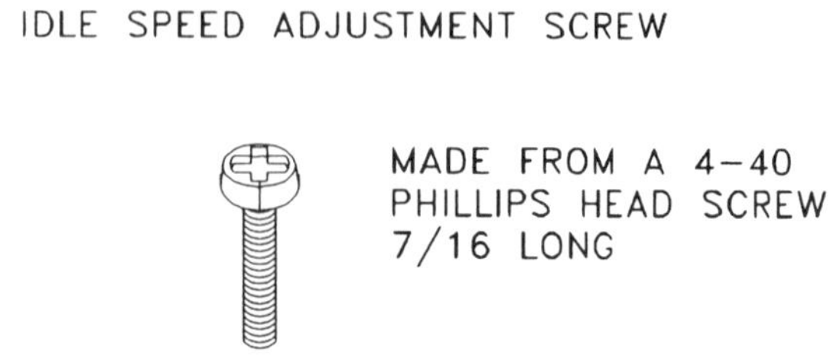

Figure 91

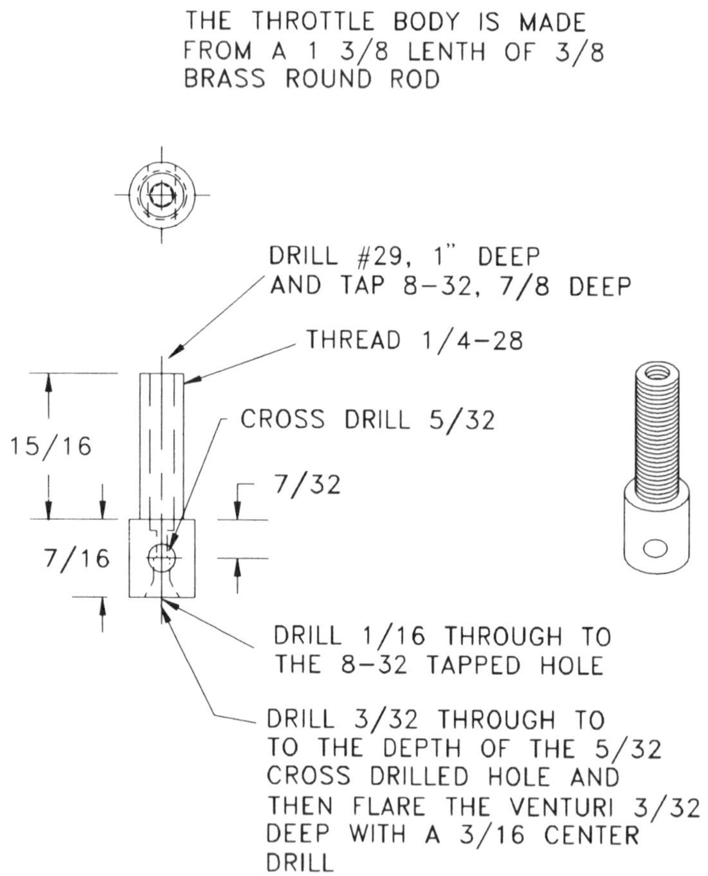

Figure 92 Throttle body

MAKING THE CARBURETOR: When we began building the engine, the carburetor seemed to be one the most intimidating things we had to look forward to. Much to our surprise building it turned out to be the most exciting and fun part of the project. An amazing and important lesson that I have learned, is that often those things that seem impossible really aren't. You simply look at a part or thing that you want to make, decide that you are going to make it despite the perceived difficulty and imaginary road blocks and begin. The problems just seem to melt away. What seemed impossible the day before is now possible and the satisfaction gained from overcoming these small hurdles makes it all worthwhile.

The carburetor Jim chose to use on his engine is a throttle barrel type. It is similar to the types of carburetors used on model airplane and model boat engines. Having subscribed to Home Shop Machinist magazine for many years we happened to notice that the carburetor Jim used is a near copy to one Bill Reichart used on his Panther Pup engine project described in the May/June 1990 issue of the magazine. The shape of the carburetor is a little different and there are some differences in drill and tap sizes and the fuel inlet is totally different, but the overall design concept is the same.

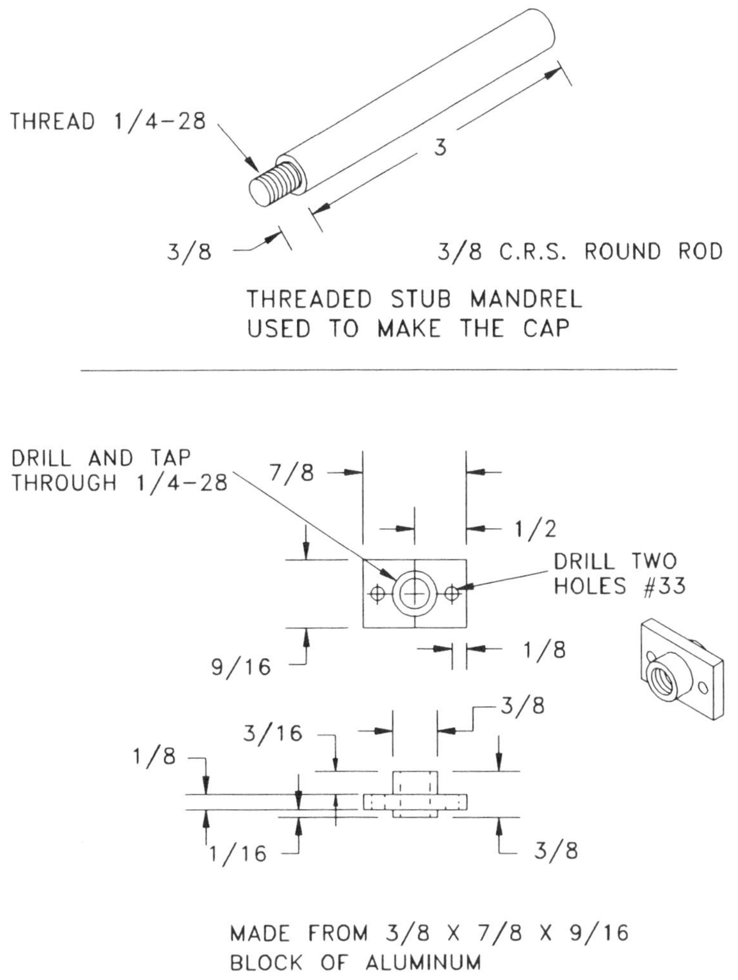

Figure 93 **Carburetor cap**

The carburetor as detailed is made of aluminum, but brass would work as well. See figure 90 for a drawing showing dimensions for making the carburetor body.

Begin by squaring up a block of aluminum 9/16" x 1" x 1-1/2". Clamp and center a 9/16" x 1" end of the carburetor blank in a 4-jaw chuck. Drill through with a #29 drill. Enlarge the hole to 13/64", 3/4" deep. Enlarge to 23/64", 5/8" deep. Ream .376", 5/8" deep. Tap the remaining portion of the #29 hole 8-32.

Reposition the carburetor in the chuck for cross drilling into the air intake side. Drill through 5/32". Enlarge to 7/32" 5/16" deep. Flare the end 1/8" deep with a 45 degree countersink.

Remove and rechuck the blank to enlarge the outlet end to 7/32", 5/16" deep. Turn the nipple end to 5/16" diameter 5/16" back. Thread nipple 5/16-24. Remove the blank from the chuck and saw out the corner on the inlet end as shown.

Mark off the location for the idle air screw and drill #54 to hit the 13/64" hole. Enlarge the hole with a #43 drill, 7/32" deep and tap 4-40. Mark off the location of the air bleed hole and drill with a #54 drill to intersect the 4-40 threaded hole. Cut a 4-40 screw to 7/16" length to make the idle air screw. Place a 1/2" long piece of a ball point pen spring over the screw and install it in the idle air hole.

The throttle body is shown in figure 92. You'll need a 3/8" collet adapter to hold the 3/8" x 1-3/8" brass round rod that the throttle barrel is made out of. Turn the blank to 1/4" diameter 15/16" back. Thread 1/4-28. Center drill and then drill #29, 1" deep. Tap 8-32 7/8" deep. Drill 1/16", 1-7/32" deep.

Remove the blank from the chuck and cross drill the 1/8" hole at the center of the barrel. Rechuck and drill 3/32" from the barrel end to the 1/8" cross drilled hole. Flare the venturi 3/32" deep with a 3/16"

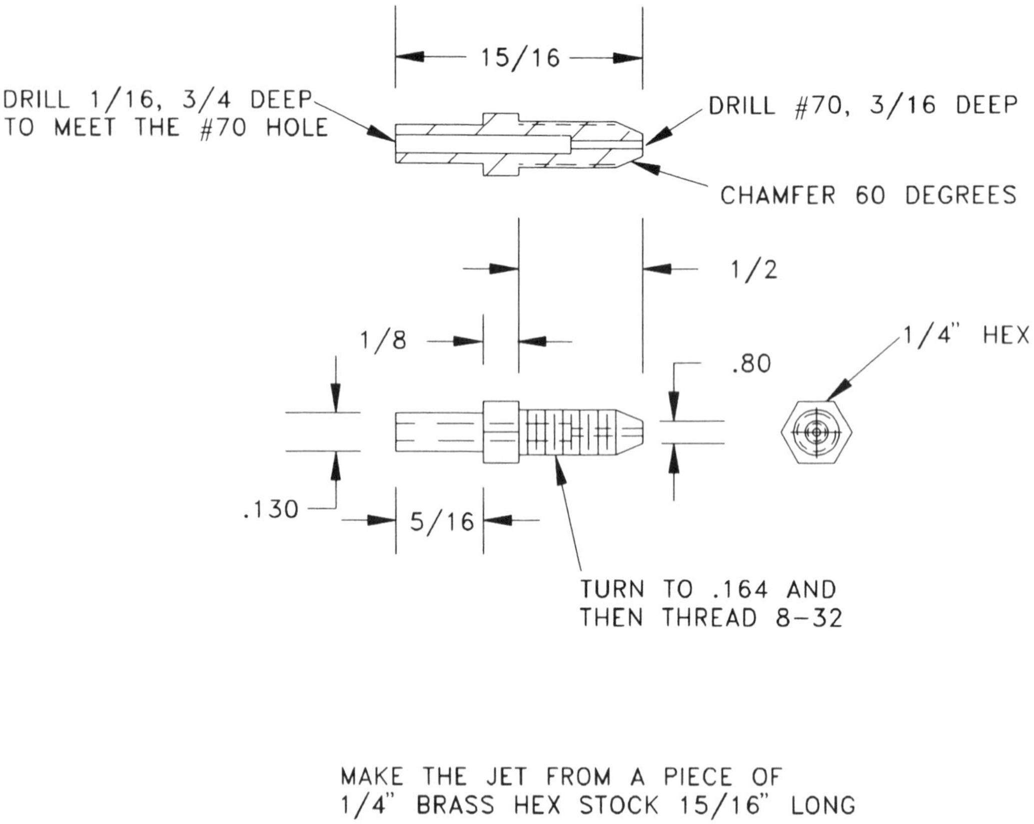

Figure 94 **The jet**

center drill.

Make the cap for the carburetor from a piece of aluminum 3/8" x 7/8" x 9/16". To make the cap you will need a stub mandrel approximately 3" long and threaded 1/4-28, 3/8" back.

Drill and tap through the cap 1/4-28 as shown in figure 93. Mount the blank on the mandrel in the lathe. Machine the 3/8" x 1/16" guide boss. Reverse on the mandrel and machine the outside 3/8" x 3/16" boss. Screw throttle into the cap and assemble to the carburetor body. Mark the location for the retainer holes and drill 3/32" through the cap and into the body to the air passage. Enlarge the holes in the cap to #33. Tap the holes in the carburetor body 4-40 and install screws 1/4" long to attach the cap to the carburetor body.

Align the throttle with the air inlet by inserting a 1/8" drill. Lock the throttle to the cap with a 1/4-28 nut. Progressively enlarge the hole in the throttle to 5/32.

Make the jet as shown in figure 94 from a piece 1/4" x 15/16" brass hex rod. Place it in a 3-jaw chuck and turn to .164", 1/2" back. Spot center and drill #70, 3/16" deep. Machine the end to a 60 degree angle leaving a slight flat for the #70 hole. Thread 8-32 to the shoulder.

Prepare a chuck adapter threaded 8-32. Screw the jet into the adapter. Spot center and drill 1/16", 3/4" deep to meet the #70 hole. Turn the end to .130", 5/16" back leaving a 1/8" wide hex shoulder. Install the jet in the carburetor body.

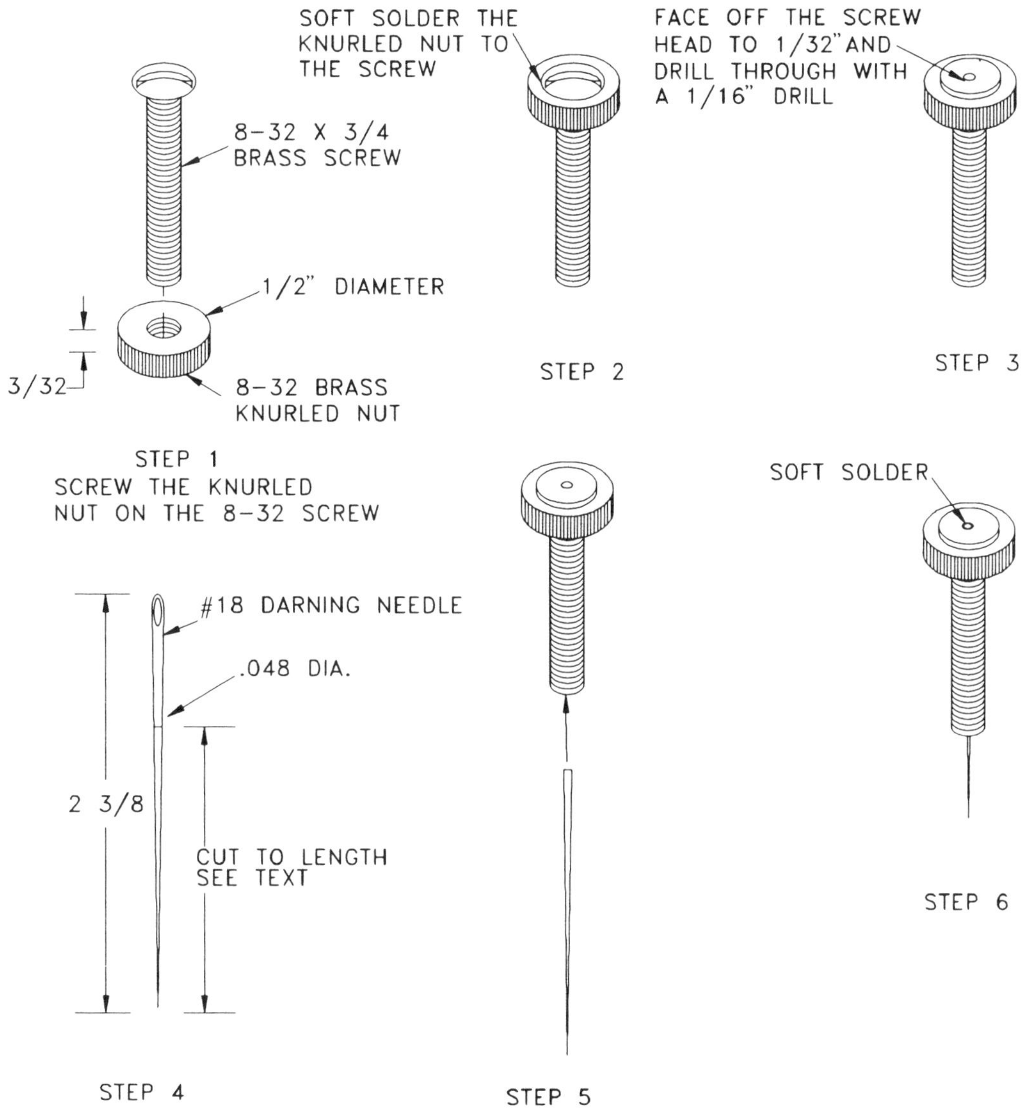

Figure 95 Needle valve

The needle valve is made from an 8-32 x 3/4" brass screw, a #18 darning needle and a knurled brass nut. See figure 95.

To make the knurled nut, chuck a piece of 1/2" brass round in the lathe. Apply a single straight knurl, medium or coarse. Drill and tap 8-32. Part off the knurled disc 3/32" thick.

Thread the knurled nut on the 8-32 x 3/4" brass screw. Soft solder the knurled disc (thumb wheel) to the screw head. Mount the screw in the 8-32 threaded adapter and face off the head of the screw to 1/32". Spot center and drill 1/16" through the screw. Screw the needle valve screw into the throttle and back out 3 turns. Slip a #18 darning needle in to seat in the jet. Mark the needle with tape where it protrudes from the screw. Score the needle on the grinder and break off the excess

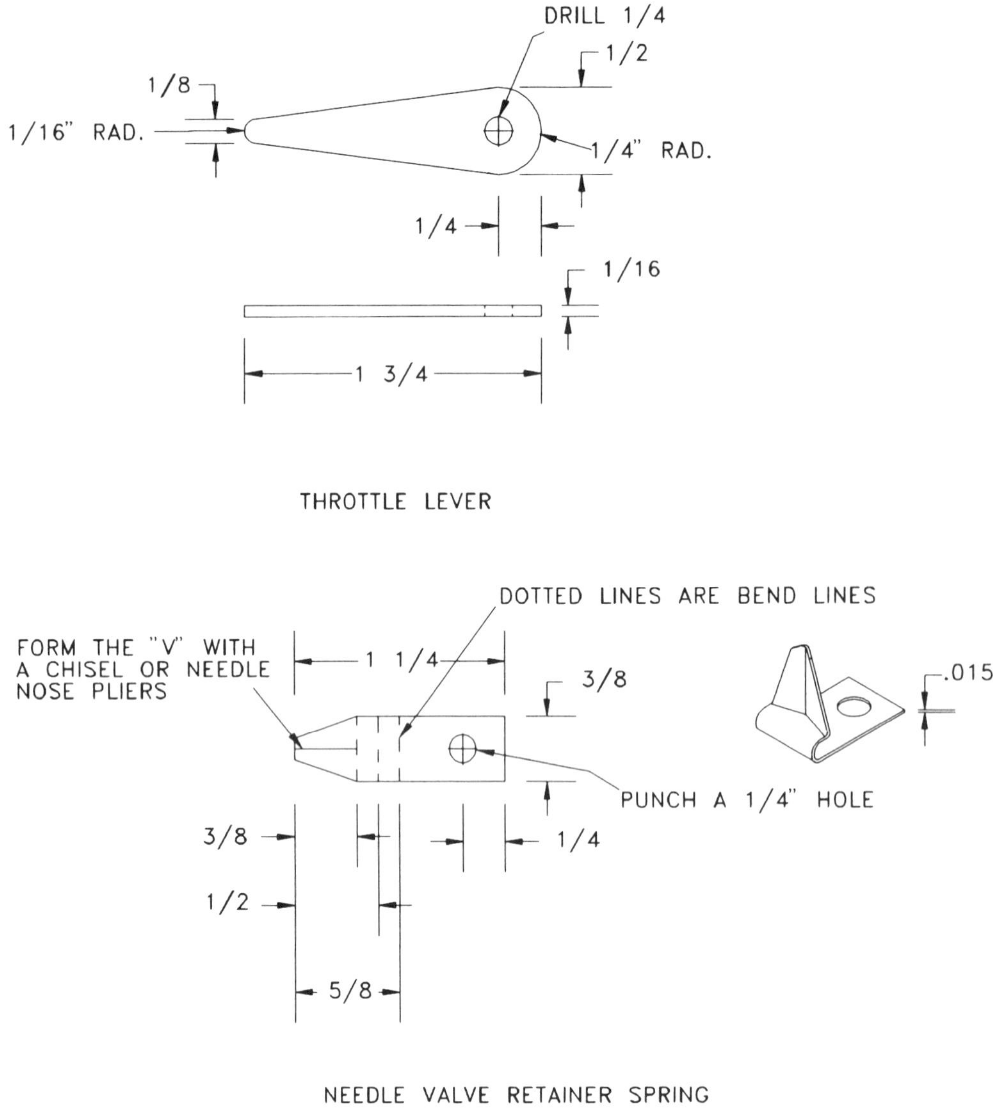

Figure 96

length. There is a varnish coating on sewing needles that prevents solder from sticking to them. To remove the coating rough up the solder end of the needle with a piece of sandpaper. Reinstall needle as shown in step 5 in figure 95. Apply flux and soft solder around the needle and into the head of the screw.

The throttle lever and the needle valve spring are the final items to make for the carburetor. They are shown in figure 96. The throttle lever is made from 1/16" sheet steel and is so simple to make that it really doesn't need discussion. Make the needle valve spring by heating a piece of clock spring from .01 to .020 thick to a bright red and cool it slowly to anneal it. Lay out the spring. Drilling thin spring steel is difficult so it is better to punch the 1/4" hole, then cut to shape. Form the V-shape in the small end with needle nose pliers. Bend to form and install to test. Remove and reharden by

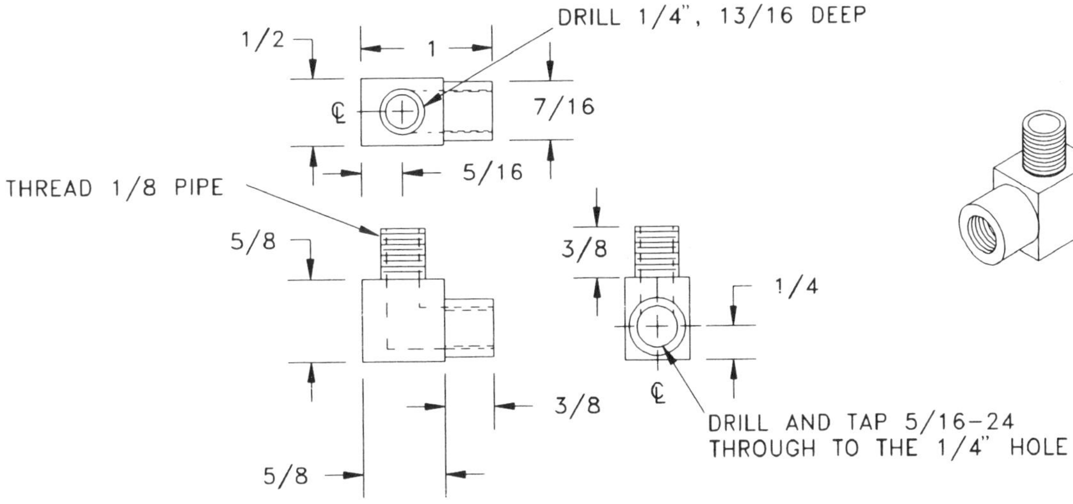

Figure 97 **Connecting elbow for the carburetor**

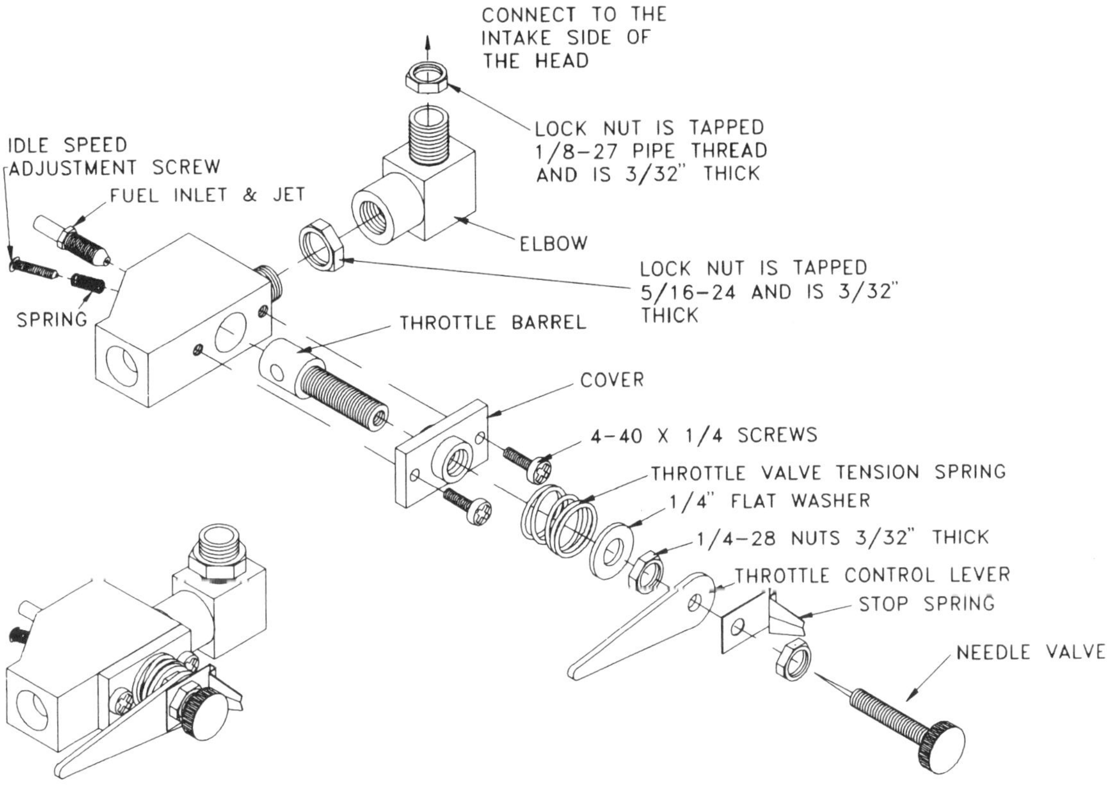

Figure 98 **Carburetor**

Figure 99 Carburetor

heating to a bright red and quenching in oil or water. Polish with emery cloth and reheat to a blue color and quench.

An elbow fitting is needed to attach the carburetor to the intake side of the head. You can make the elbow from a 1/2" x 1" x 1" block of aluminum or brass. Figure 97 shows the dimensions. Place the block in a 4-jaw chuck and drill the 1/4" hole 13/16" deep. Turn the end to 3/8", 3/8" back and thread for 1/8" pipe thread. Rechuck the block and Drill and tap 5/16-24 through to the 1/8" pipe threaded hole. Turn the O.D. of the end to 7/16", 3/8" back.

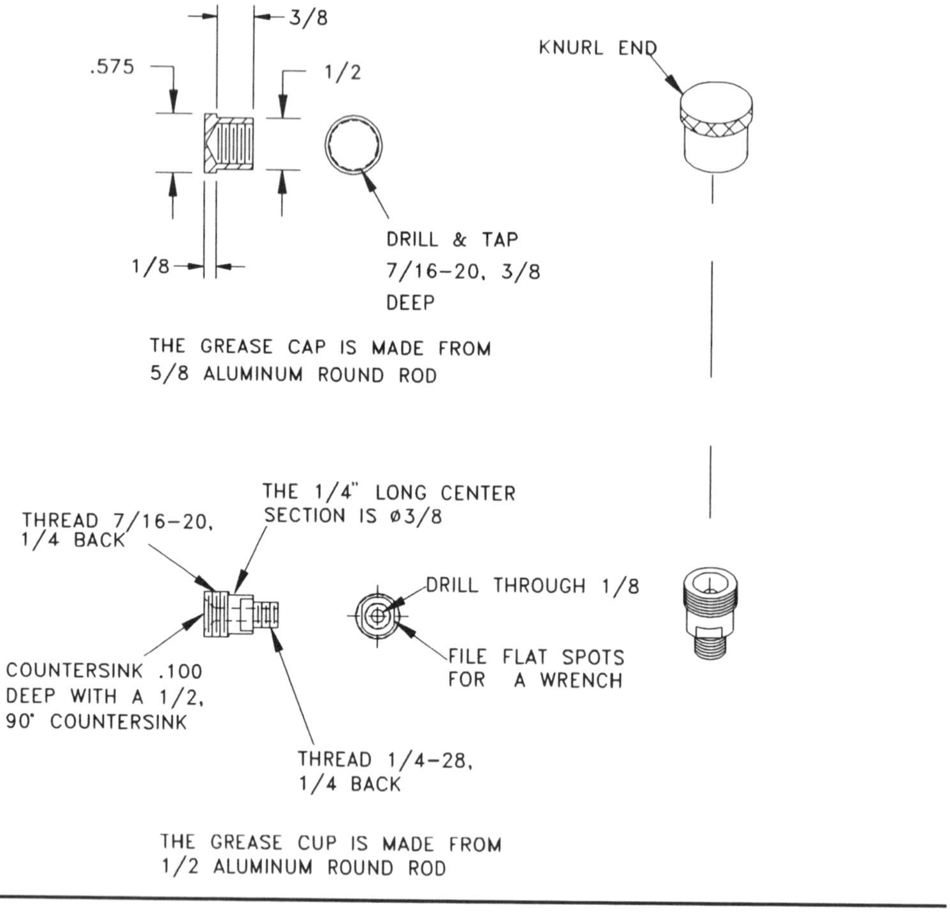

Figure 100 Grease cups. Make three

Figure 98 shows an exploded view of the carburetor assembly. Two 1/4" nuts act as jam nuts and tighten against both sides of the throttle control lever keeping it locked on the shaft. The throttle valve tension spring is something we just happened to have laying around. It measures 1/2" OD x 1/2" long. The wire thickness of the spring is .045". A 1/4" flat washer goes between the spring and the throttle lever lock nut.

You will have to remove the head from the water jacket to attach the elbow and carburetor to the intake connection. Be sure and use locknuts on both ends of the elbow to keep the carburetor connection from loosening while the engine is operating.

MAKING THE GREASE CUPS: The crankshaft and the link pin in the pivot post are lubricated by grease cups. These work by putting grease in the cap portion of the fitting and as the cap is tightened grease is forced onto the shaft. The grease cap is made from 5/8" aluminum round rod and the cup is made from 1/2" aluminum round rod. Dimensions are shown in figure 100.

To make the cap, chuck a length of 5/8" aluminum round rod in a 3-jaw chuck so that about 1" sticks out past the face of the chuck. Face off, drill and tap 7/16-20, 3/8" deep. Turn the 1/2" shoulder 3/8" back from the end. Knurl the 1/8" section and part off at 1/2" in length. Repeat for the next two caps.

A 1/4-28 threaded adapter is needed to make the cup. To make the cup, chuck a short piece of 1/2" aluminum round rod in a 3-jaw chuck so that 1" sticks out past the chuck. Reduce to 3/8" diameter 1/2" from the end. Reduce to 1/4" diameter 1/4" from the end. Thread 1/4-28 using a tail stock die holder. File flats for a wrench as shown. Drill 1/8", 7/8" deep. Part off at 3/4". Repeat for two more cups. Mount the prepared blanks in the 1/4-28 adapter to face off, counter sink and thread 7/16-20.

When the cups are made screw one into each of the crankshaft support pillars and one into the pivot post. It is possible that the 1/4-28 threaded end could be too long and bind against the crankshaft when the cup is tightened. If that happens file the end a little to cure the problem.

Figure 101 Gas tank

THE GAS TANK: The gas tank was made from stuff we had laying around the shop. A 2" long, 1-3/4" ID piece of exhaust pipe forms the main body and 1-7/8" freeze plugs (the type used in engine blocks) serve for the ends. The inlet and outlet fittings are made from brass round rod. The legs are made from 1/16" sheet steel.

The drawings for the gas tank are shown in figures 102 through 104. Make the center section first. It is important when welding or brazing a tank to allow a way for the welding gases to escape or you could create a miniature bomb, so drill the holes in the center section first before brazing the ends on. After the holes are

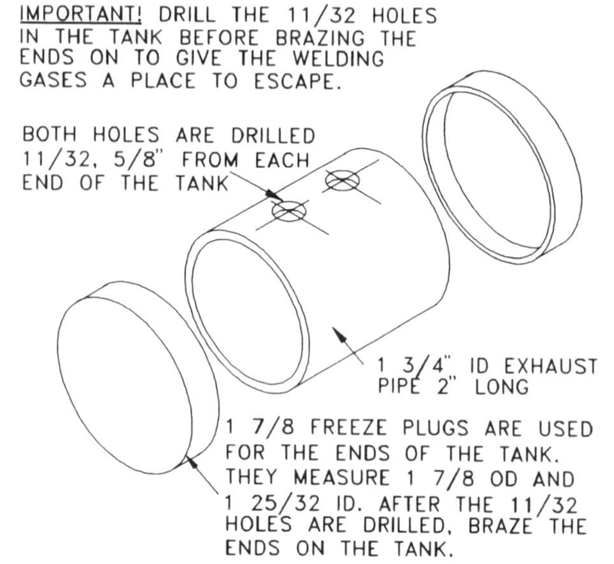

Figure 102

drilled, braze the ends on.

Make the legs for the gas tank as shown in figure 103. They are made out of 1/16" sheet steel. Braze them to the tank just behind the lip of each end cap.

Make the inlet and outlet fittings. The gas cap is made from 1/2" aluminum round rod using the same methods used to make the cap for the grease cup. Be sure to drill the vent hole in the top of the gas cap with a #60 drill.

The inlet fitting is made from 3/8" brass round rod drilled through 17/64" and threaded on the outside 3/8-24, 7/16" back. Turn a .340" diameter lip on the unthreaded end to fit the drilled hole in the tank.

The outlet fitting is made from 3/8" brass round rod drilled and tapped through 5/16-24. Turn a 5/16" diameter lip on one end 1/16" back to fit the drilled hole in the tank. When the inlet and outlet fittings are made, soft solder them to the tank. When soft soldering it is best to tin the area to be soldered together. Tinning means applying a thin coat of solder to each surface before actually soldering the parts together. This makes the soldering job easier and makes a much better joint. The fuel outlet adapter is made from 5/16" brass round rod. Thread one end 5/16-24, 5/16" back. Drill through 1/8", then file a flat spot on the unthreaded end to fit a 1/4" wrench. Screw the outlet adapter into the fuel outlet fitting. Insert a 2-1/4" length of 1/8" brass tubing with one end trimmed at a 45 degree angle. Allow the angled end to rest against the bottom of the tank and soft solder the tube to the fitting.

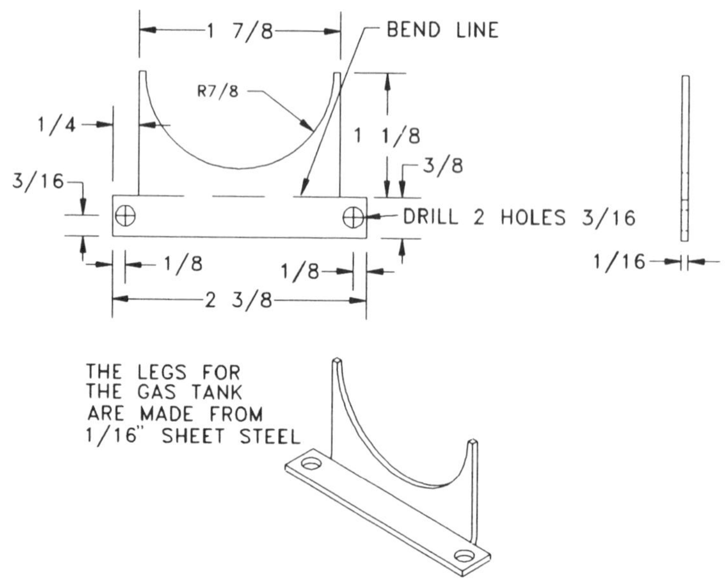

Figure 103 Gas tank legs

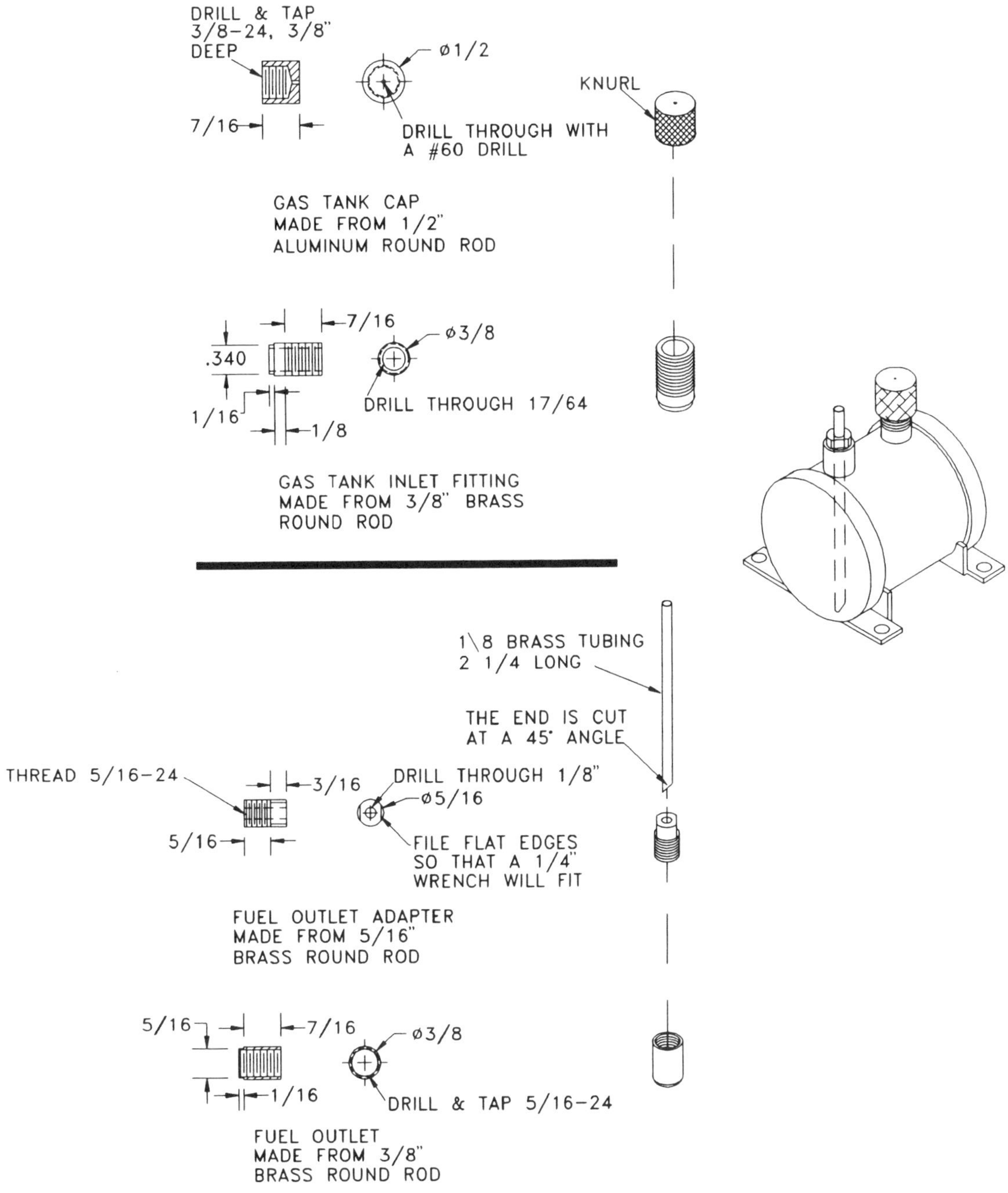

Figure 104 Gas tank fittings

THE CYLINDER OILER: We used a 1" x 1/8" snap lever sight feed drip oiler to lubricate the cylinder. These are available at a reasonable cost from several mail order sources. They have a clear acrylic lens and an on/off lever on top with a knurled adjustable feed knob below the top of the switch. We purchased ours from Wholesale Tool, 12155 Stephens Drive, Box 68, Warren, Michigan 48090, Phone 1-800-521-3420.

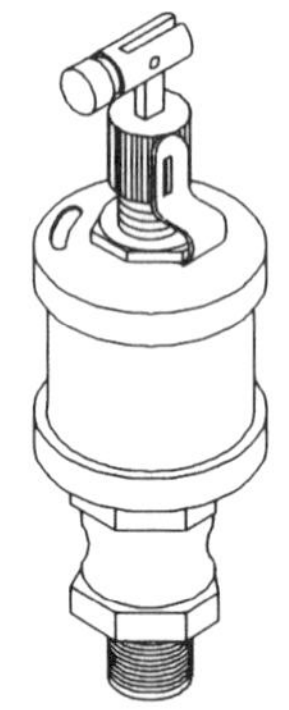

THE SPARK PLUG: There are a couple of sparkplugs that will work in this application. They are an NGK #CM-6 or a Champion #Y82. These are small plugs with a 10MM-1.0 thread. They can be purchased at automotive supply stores and motorcycle parts stores.

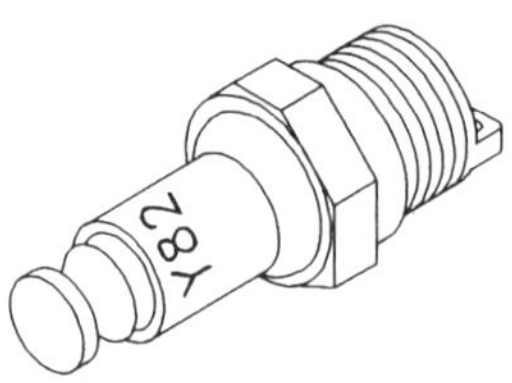

THE IGNITION POINTS: The ignition points and condenser come from a late 1970's Ford V-8 engine. They are available at most auto parts stores and I even found them at our local Wal-Mart store. Figure 107 shows them being mounted to the ignition plate. Two 8-32 x 1/2" screws hold the points in place and a 6-32 x 1/2" screw fastens the condenser mounting bracket.

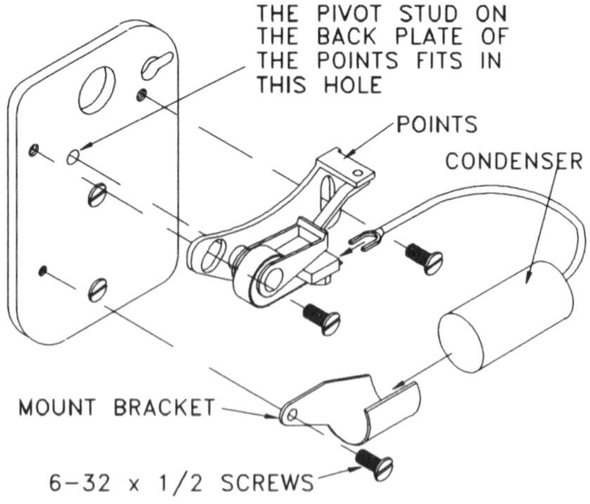

Figure 107 **Mounting the points**

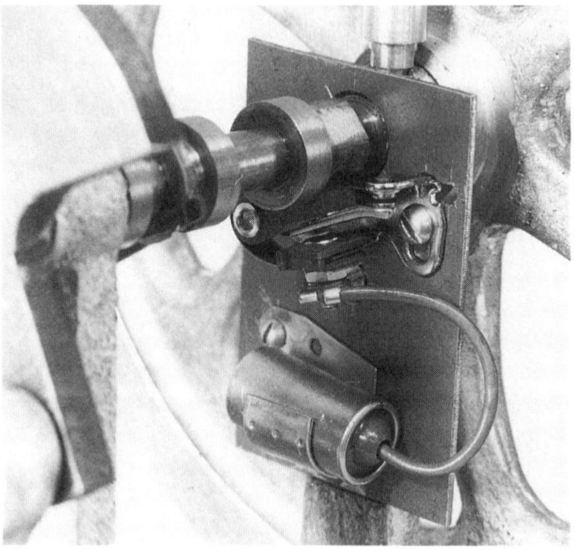

Figure 108

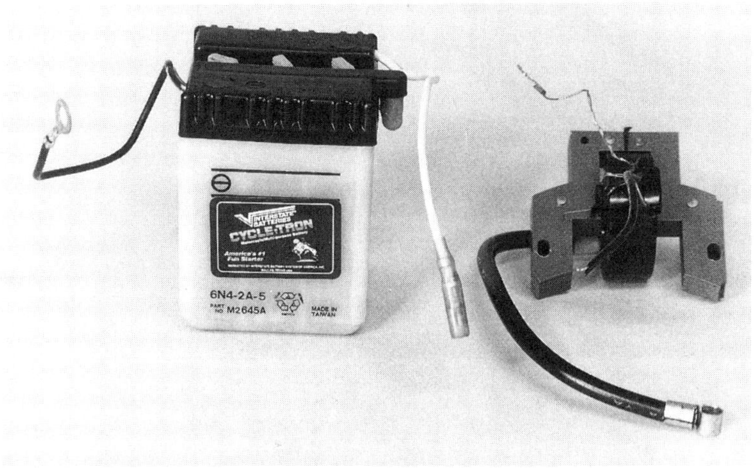

Figure 109 The battery and coil

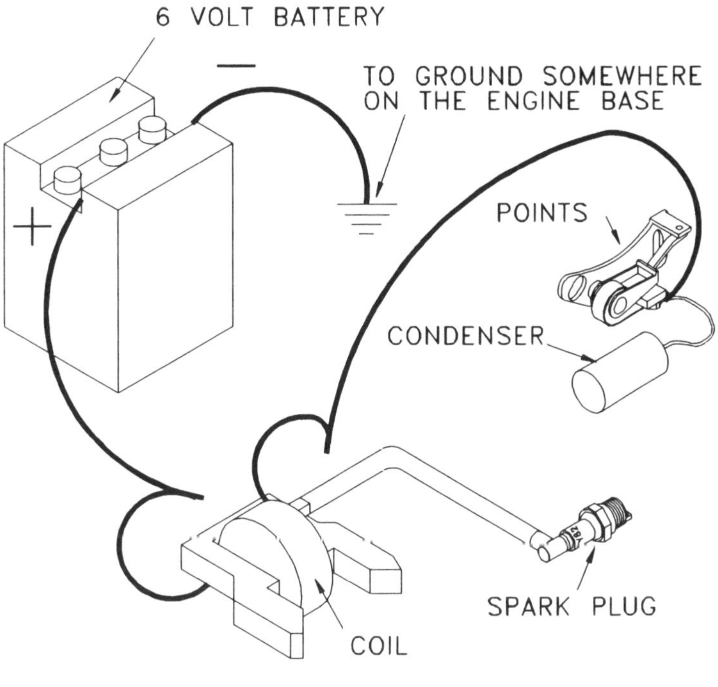

Figure 110 The wiring diagram

THE ELECTRICAL SYSTEM: The battery used measures 2-3/4" x 2-3/4" x 3-3/4". These smaller types of rechargeable batteries are sold at motorcycle parts stores. The coil is the type used on small Briggs and Stratton engines. See figure 109 for a photo of the coil and battery. The wiring for the engine is simple and a schematic is shown in figure 110. The positive terminal of the battery is connected to one of the coil leads and the other coil lead is connected to the points. The negative terminal of the battery is grounded to the base of the engine. And of course the spark plug wire is connected to the spark plug.

SETTING THE CAMS AND ADJUSTING THE POINTS: There is a myth floating around that says the Atkinson "Cycle" engine was a nightmare to time. This is not the case, you will be amazed at how simple it is.

To begin refer to figures 5 through 8 at the beginning of the book. Relax, pull up a chair and have a seat in front of the engine so you can have an eye level view of it. Slowly turn the flywheel through a few revolutions and familiarize yourself with the position of the crank as it travels through each stroke.

Use the drawings to make comparisons.

Adjust the intake cam first. At the beginning of the intake stroke the exhaust valve has just closed and all of the exhaust gases have been removed from the cylinder. The object is to set the intake valve to open at the exact time the piston reaches the top of its forward stroke and starts to move back away from the head. Adjust the cam to accomplish this as shown in figure 5 and tighten the set screw.

The compression stroke shown in figure 6 follows the intake stroke. During this stroke both of the valves are closed and the vapored fuel is being compressed in the top of the cylinder.

Adjust the points to break when the piston reaches the top of the compression stroke. When the points break contact a spark is created in the cylinder that ignites the compressed fuel and forces the piston into the power stroke shown in figure 7.

As the piston reaches the back of the power stroke it begins to move forward to exhaust the spent vapors. Adjust the exhaust cam to force the exhaust valve open at this

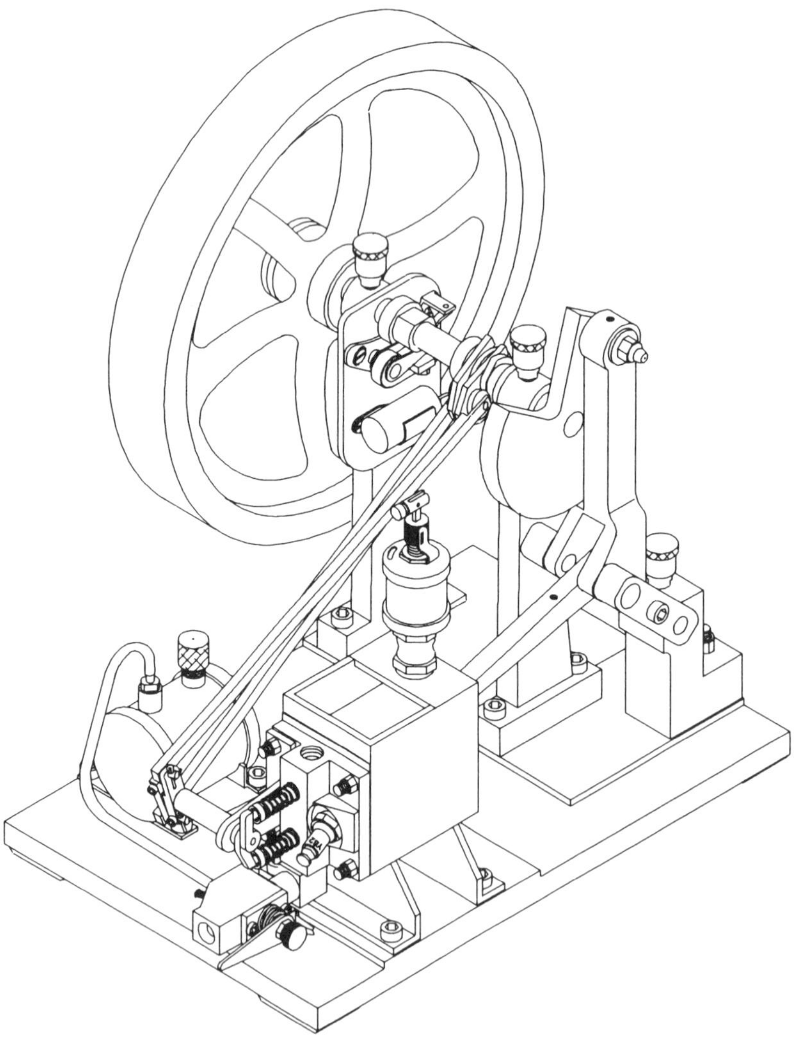

Figure 111 The engine ready to start

point as shown in figure 8 and tighten the set screw.

That's all there is to it. Turn the flywheel through a couple of revolutions to check your adjustments. The shorter stroke is the intake stroke so it is important that the exhaust valve be totally closed at the beginning of this stroke and that the intake valve be open to allow as much fuel as possible into the cylinder. It is very important that both valves be closed during the compression stroke to prevent fuel from being wasted into the atmosphere before it can be used.

FINAL PREPARATION: No doubt you are anxious to give the flywheel a spin and find out if all your hard work has paid off. First, give yourself a break, get yourself a cup of coffee or whatever else it is you do in situations like this, and give yourself a big pat on the back for a job well done because chances are in a few minutes you're going to have a running engine.

Before giving it a try, position the gas tank in the empty area on the base to the left of the water jacket. Mark the hole locations corresponding with the mounting holes in the legs of the gas tank. Drill and tap them 8-32. Fasten the gas tank to the base with 8-32 x 3/8" socket head cap screws. Connect the gas tank to the carburetor with a short piece of 1/8" clear vinyl hose. Fill the grease cups with grease and oil all of the moving parts. Don't forget to squirt a little oil in the cylinder and on the piston wrist pin. Fill the cylinder oiler with 30 weight oil. Clamp the engine to a table with a C-clamp. Fill the water jacket with antifreeze solution. Hook up the wiring and battery as shown in figure 110. Short lengths of 14 ga. wire with alligator clips on the end work well for a temporary set up. Make sure the coil is not laying on anything metal. I first clamped my engine to my welding table with the coil laying on it. I spent several frustrating hours trying to get the engine started before I discovered that the metal table top was interfering with the magnetic field of the coil.

Just before spinning the flywheel for the first time adjust the idle air screw and needle valve screw out about 1/2 turn. Put about six drops of oil into the exhaust opening in the top of the head. This lubricates the top of the cylinder. It may be necessary to add oil to the top of the cylinder for the first several times you run the engine until it breaks in.

You're ready to give it a try. Spin the flywheel and see what happens. Chances are it'll start right up. If necessary you can choke the carburetor by placing your finger over the air intake and then removing it when the engine fires. Open the needle valve slightly each time you spin the flywheel. It shouldn't take long before your engine starts. If it doesn't start after a few tries there is more than likely a problem somewhere. First make sure that gas is getting sucked to the carburetor. If not, there are a couple of things that could prevent it from getting there. There could be a piece of dirt in the carburetor. If that's not it, check to make sure your cams are opening and closing the valves in the right sequence. If that looks okay then you probably have a compression problem. On all four of the engines we built we had a piece of metal shaving lodge in the valve seat and cause a loss of compression. This happened regardless of how well we cleaned the head and carburetor before attempting to start the engine. This usually happened after the engine had run a few minutes. But anyway if you have a

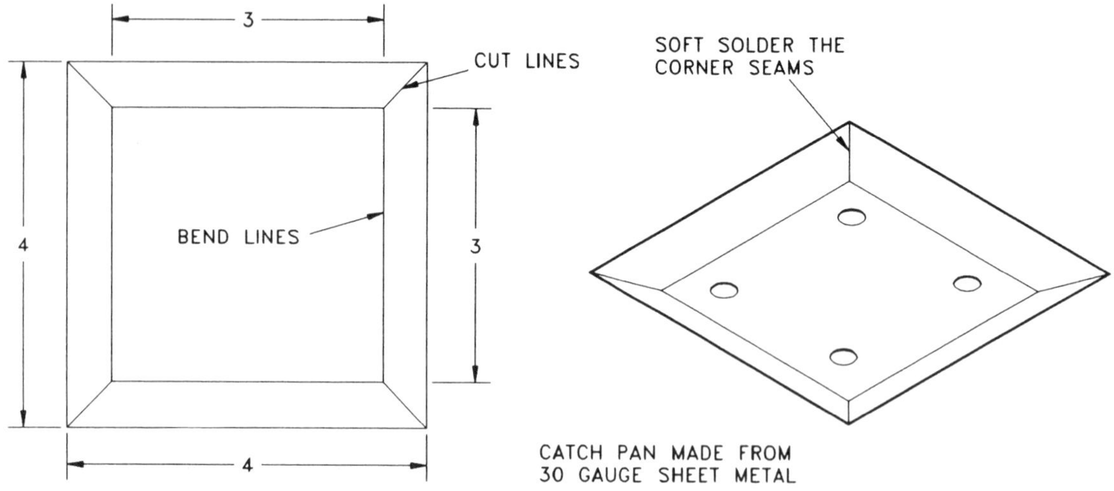

Figure 112 **Catch pan**

compression problem at any time and you are sure you lapped the valves properly it is probably caused by a piece of dirt in the valve seat. The solution would be to remove the head, clean the valve seats and give the engine another try. If you are getting fuel to the carburetor and the engine still doesn't run, check to see if the points are making contact and breaking when the piston is at top center. Also check to see if the points are firing and that there is a spark at the plug. Make sure the wiring is correct and that the battery is fully charged.

Your engine will take several running sessions to finally break in. During the breakin period, it will perform a little better each time you run it. And it won't hurt to experiment by changing the cam and point adjustments slightly, then checking for improved performance. You will find the amazing thing about the "Cycle" engine is it's reliability and easy starting. It runs amazingly smooth even at very slow idle speeds.

SPRUCING UP AND ADDING A FEW EXTRAS: One added accessory is an exhaust stack. It quiets the exhaust noise down and prevents the antifreeze in the water jacket from splashing into the exhaust port in the head. Make it 1-1/2" long out of a 1/8" brass pipe nipple. For a better appearance you can cut the top off at a 45 degree angle.

You will discover that oil from the cylinder has a tendency to drip out and run across the base. You may decide to make a catch pan and mount it under the water jacket. It'll make cleanup a little easier and adds an extra touch to the appearance of the engine. We made ours out of sheet metal. Cut piece of 30 gauge sheet metal to size as shown in figure 112. Cut it diagonally at the corners and bend all four ends up at a slight angle and solder the corner seams. Holes are drilled in the pan to align with the holes in the mounting legs of the water jacket.

Painting the engine is another important

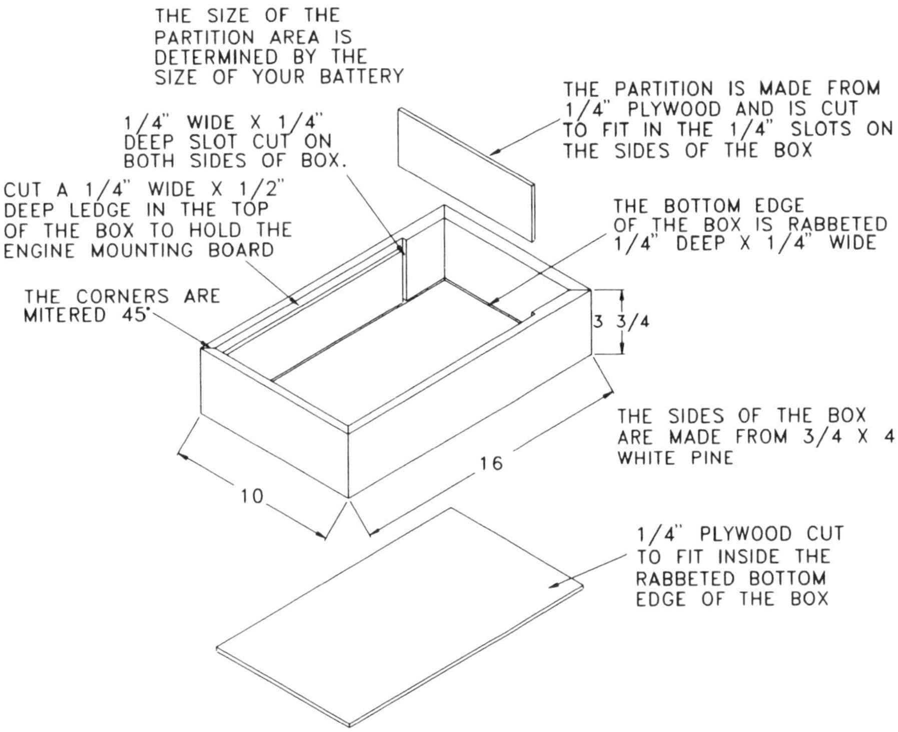

Figure 113

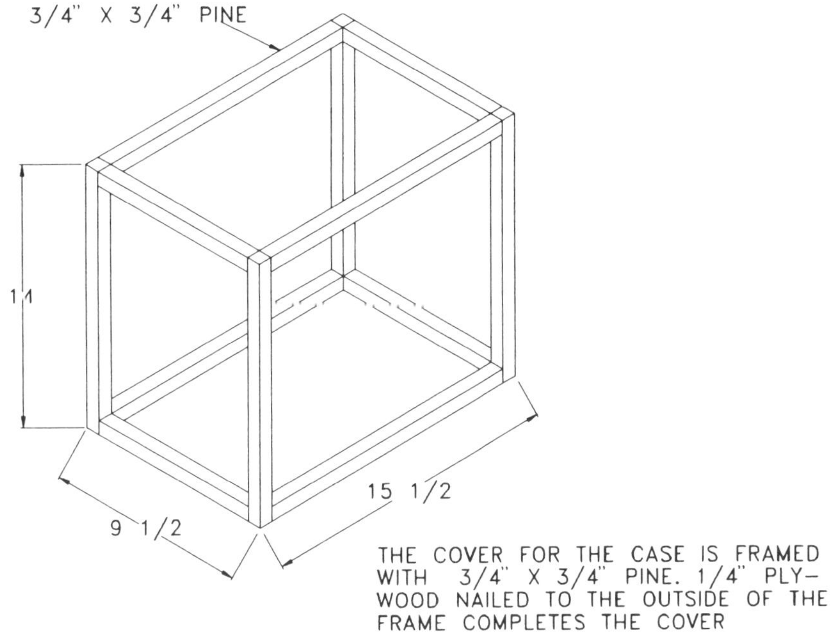

Figure 114

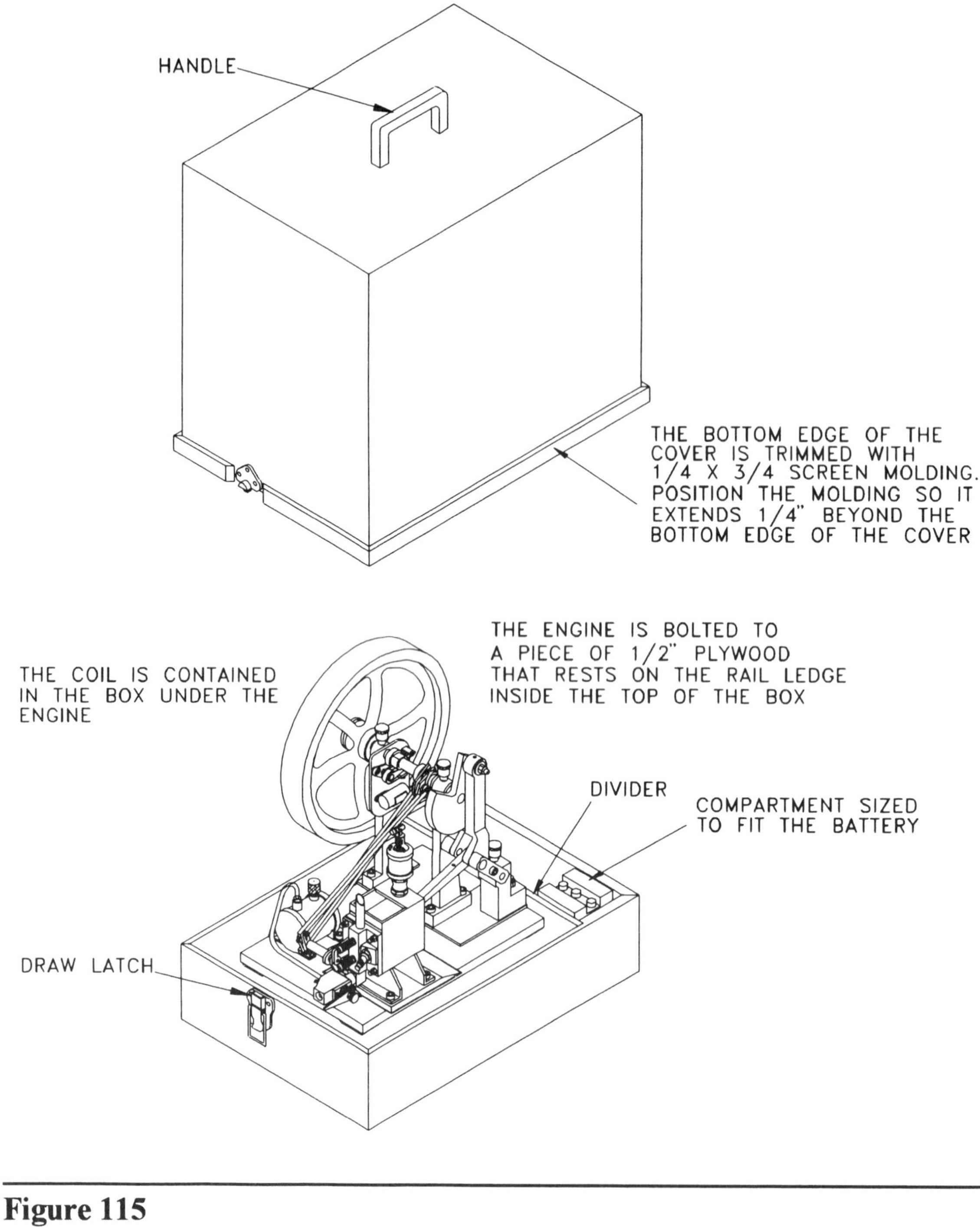

Figure 115

thing to consider. The original colors on the Atkinson engine were dark green with a dark red crank throw so we stuck with those. I painted the water jacket, inside spokes of the flywheel and the gas tank dark green. The crank throw was painted a dark red and the ignition plate and rocker arm assembly were painted black.

Chances are you will want to take your engine with you when you go visiting and show it off. We made a nice case for ours patterned after a portable sewing machine case. It is made a little longer than the engine to make room for a compartment for the battery. The cover is sized to fit over the engine and is secured with draw catch latches at each end. A handle is mounted on the top. If you're a good carpenter you can use high quality material and stain and varnish the case. Otherwise use cheaper material and fill the imperfections with wood filler and paint it.

A drawing of the lower portion of the case is shown in figure 113. It is made from 3/4" x 4 white pine. The corners are mitered and the sides are glued and nailed together. A 1/4" wide x 1/2" deep ledge is cut along the inside top edge of the box. A 1/4" wide x 1/4" deep slot is cut on each side of the box to hold the 1/4" plywood partition that forms the battery storage area. The size of your battery will determine the location of the partition divider. The bottom edge of the case is rabbeted 1/4" wide and 1/4" deep. The 1/4" plywood bottom is cut to fit inside the rabbeted area and then glued and nailed in place.

The cover portion of the case is shown in figure 114. It is framed with 3/4" x 3/4" white pine and is faced with 1/4" plywood nailed to the outside. The bottom edge of the cover is trimmed with 1/4" x 3/4" screen molding as shown in figure 115. The molding is positioned so that 1/4" extends below the bottom edge of the cover. This is so that when the case is put together the molding sets along the side and below the top edge of the bottom section preventing the cover from moving from side to side.

The engine is bolted to a 1/2" plywood board that is cut to fit in the top of the bottom section of the case. The board will rest on the 1/4" wide x 1/2" deep ledge. The coil can be mounted inside the box and be hidden from view by the engine mounting board. The wiring is neatly routed to the battery, points and spark plug.

Congratulations on a job well done. There is nothing to compare with the satisfaction gained from building an engine and then having it actually run. I can assure you, very few people have a running version of an Atkinson "Cycle" engine, let alone one they've built themselves. It's something to be proud of and you'll really enjoy showing it off to your friends.